Environmental Education

Published by:
Saurabh Publishing House

Environmental Education

Dr. A.L. Joshi

Saurabh
Publishing House
4735/22, Prakash Deep Building,
Ansari Road, Daryaganj
New Delhi- 110002

Published by :
SAURABH PUBLISHING HOUSE

Distributed by :
Lotus Press : Publishers & Distributors
Unit No. 220, 2nd Floor, 4735/22, Prakash Deep Building,
Ansari Road, Darya Ganj, New Delhi- 110002
Ph.: 32903912, 23280047, 09811838000
• E-mail : lotus_press@sify.com
www.lotuspress.co.in

Saurabh Publishing is an imprint of
Lotus Press : Publishers & Distributors

Environmental Education

ISBN: 978-81-89005-95-5 (H/B)

Printed at : Concept Imprint, Delhi

Preface

The Indian Constitution laid down the responsibility of Government to protect and improve the environment and made it a "fundamental duty of every citizen to protect and improve the natural environment including forests, lakes, rivers and wildlife". On this background Department of Environment was established by the Government of India in 1980 and a Ministry was formed in 1985. The Constitution and the Government's commitment to the environment along with the environmentally sound practices is an important backdrop under which the Environment Education (EE) strategy has been evolved.

The GLOBE is an International Science and Education Programme, which emphasizes on hands-on participatory approach. India joined this programme in August, 2000. This programme unites students, teachers and scientists all over the world and targets school children. The students of GLOBE schools are required to collect data about various basic environmental parametres under the supervision of a GLOBE trained teacher. Through this they learn about scientific protocols and perform environmental learning activities, which have already been introduced as theory in the textbooks. The GLOBE programme not only helps the students to appreciate the contents of the textbooks through better understanding but also assists them in gaining complete knowledge of environment. It facilitates research through a worldwide research team comprising of students, teachers and scientists.

The major topics dealt in this book are: *Introduction to Environmental Education; Strategies in Environmental Education; Challenges in India; Ecology and Science of Earth; Atmospheric Environment; Climate and Changes; Atmosphere of Earth; Environmental Pollution; Ecological Process and Sustainability; Environmentalism and Pollution Prevention; Ozone Depletion and Global Warming;The Environmental Protection Act 1986* etc.

I express my gratitude to various scholars, teachers and friends for their assistance and guidance. Finally, I thank the publisher for bringing out this book in very limited time.

– ***Author***

Contents

1
Introduction to Environmental Education

Environmental education ("EE") refers to organized efforts to teach about how natural environments function and, particularly, how human beings can manage their behaviour and ecosystems in order to live sustainably. The term is often used to imply education within the school system. However, it is sometimes used more broadly to include all efforts to educate the public and other audiences, including print materials, websites, media campaigns, etc. Related disciplines include outdoor education and experiential education.

History

The roots of environmental education can be traced back as early as the 18th century when Jean-Jacques Rousseau stressed the importance of an education that focuses on the environment in Emile, or On Education. Several decades later, Louis Agassiz, a Swiss-born naturalist, echoed Rousseau's philosophy as he encouraged students to "Study nature, not books." These two influential scholars helped lay the foundation for a concrete environmental education program, known as Nature Study, which took place in the late 19th century and early 20th century.

The nature study movement used fables and moral lessons to help students develop an appreciation of nature and embrace the natural world. Anna Botsford Cornstock, the head of the Department of Nature Study at Cornell University, was a prominent figure in the nature study movement and wrote the Handbook for Nature Study in 1911, which used nature to educate children on cultural values. Cornstock and the other leaders of the movement, such as Liberty Hyde Bailey, helped Nature Study garner tremendous amounts of support from community leaders, teachers, and scientists and change the science curriculum for children across the United States.

A new type of environmental education, Conservation Education, emerged as a result of the Great Depression and Dust Bowl during the 1920s and 1930s. Conservation Education dealt with the natural world in a drastically different way from Nature Study because it focused on rigorous scientific training rather than natural history. Conservation Education was a major scientific management and planning tool that helped solve social, economic, and environmental problems during this time period.

The modern environmental education movement, which gained significant momentum in the late 1960s and early 1970s, stems from Nature Study and Conservation Education. During this time period, many events – such as Civil Rights, the Vietnam War, and the Cold War – placed Americans at odds with one another and the U.S. government. However, as more people began to fear the fallout from radiation, the chemical pesticides mentioned in Rachel Carson's Silent Spring, and the significant amounts of air pollution and waste, the public's concern for their health and the health of their natural environment led to a unifying phenomenon known as environmentalism. Ultimately, the first Earth Day on April 22nd, 1970 – a national teach-in about environmental problems – paved the way for the modern environmental education movement. Later that same year, President Nixon passed the National Environmental Education Act. Then, in 1971, the

National Association for Environmental Education (now known as the North American Association for Environmental Education) was created to improve environmental literacy by providing resources to teachers and promoting environmental education programmes.

Internationally, environmental education gained recognition when the UN Conference on the Human Environment held in Stockholm, Sweden, in 1972, declared environmental education must be used as a tool to address global environmental problems. The United Nations Education Scientific and Cultural Organization (UNESCO) and United Nations Environment Programme (UNEP) created three major declarations that have guided the course of environmental education.

Stockholm Declaration

June 5-16 1972-The Declaration of the United Nations Conference on Human Environment. The document was made up of 7 proclamations and 26 principles "to inspire and guide the peoples of the world in the preservation and enhancement of the human environment.

The Belgrade Charter

October 13-22 1975-The Belgrade Charter was the outcome of the International Workshop on Environmental Education held in Belgrade, Yugoslavia. The Belgrade Charter was built upon the Stockholm Declaration and adds goals, objectives, and guiding principles of environmental education programmes. It defines an audience for environmental education, which includes the general public.

The Tbilisi Declaration

October 14-26 1977-The Tbilisi Declaration "noted the unanimous accord in the important role of environmental education in the preservation and improvement of the world's environment, as well as in the sound and balanced development of the world's communities." The Tbilisi Declaration updated and clarified The Stockholm Declaration and The Belgrade Charter by including new goals, objectives, characteristics, and guiding principles of environmental education.

Later that decade, in 1977, the Intergovernmental Conference on Environmental Education in Tbilisi, Georgia emphasized the role of EE in preserving and improving the global environment and sought to provide the framework and guidelines for environmental education. The Conference laid out the role, objectives, and characteristics of environmental education, and provided several goals and principles for environmental education.

Modern Environmental Education

Following the 1970s, non-governmental organizations that focused on environmental education continued to form and grow, and the movement gained stronger political backing. A critical move forward came when in US Congress passed the National Environmental Education Act of 1990, which placed the Office of Environmental Education in the U.S. Environmental Protection Agency and allowed the EPA to create environmental education initiatives at the federal level.

Antecedents

In the United States some of the antecedents of Environmental Education were Nature Studies, Conservation Education and School Camping. Nature studies integrated academic approach with outdoor exploration. Conservation Education brought awareness to the misuse of natural resources. George Perkins Marsh discoursed on humanity's integral part of the natural world. The governmental agencies like the Forestry Service and the EPA were also pushing a conservation agenda. Conservation ideals still guide environmental education today. School Camping was exposure to the environment and use of resources outside of the classroom for educational purposes. The legacies of these antecedents are still present in the evolving arena of environmental education.

Environmental education has been considered an additional or elective subject in much of traditional K-12 curriculum. At the elementary school level, environmental education can take the form of science enrichment curriculum, natural history field trips, community service projects, and participation in outdoor science schools. EE policies assist schools and organizations in developing and improving environmental education programmes that provide citizens with an in-depth understanding of the environment. School related EE policies focus on three main components: curricula, green facilities, and training.

Schools can integrate environmental education into their curricula with sufficient funding from EE policies. This approach – known as using the "environment as an integrating context" for learning – inserts environmental education into the core subjects and thus environmental education does not take time away from other important subjects, such as art, gym, or music. In addition to funding environmental curricula in the classroom, environmental education policies allot the financial resources for hands-on, outdoor learning. These activities and lessons help address and mitigate "nature deficit disorder," as well as encourage healthier lifestyles.

Green schools, or green facility promotion, are another main component of environmental education policies. Greening school facilities cost, on average, a little less than two per cent more than creating a traditional school, but payback from these energy efficient buildings occur within only a few years. Environmental education policies help reduce the relatively small burden of the initial start-up costs for green schools. Green school policies also provide grants for modernization, renovation, or repair of older school facilities. Additionally, healthy food options are also a central aspect of green schools. These policies specifically focus on bringing freshly prepared food, made from high-quality, locally-grown ingredients into schools.

In secondary school, environmental curriculum can be a focused subject within the sciences or is a part of student interest groups or clubs. At the undergraduate and graduate level, it can be considered its own field within education, environmental studies, environmental science and policy, ecology, or human/cultural ecology programmes.

The North American Association for Environmental Education has established the following "Guidelines for Excellence" for environmental education:

1. **Fairness and accuracy:** EE materials should be fair and accurate in describing environmental problems, issues, and conditions, and in reflecting the diversity of perspectives on them. (i) Factual accuracy. (ii) Balanced presentation of differing viewpoints and theories. (iii) Openness to inquiry. (iv) Reflection of diversity.
2. **Depth:** EE materials should foster an awareness of the natural and built environment, an understanding of environmental concepts, conditions, and issues, and an awareness of the feelings, values, attitudes, and perceptions at the heart of environmental issues, as appropriate for different developmental levels. (i) Awareness. (ii) Focus on concepts. (iii) Concepts in context. (iv) Attention to different scales.
3. **Emphasis on skills building:** EE materials should build lifelong skills that enable learners to address environmental issues. (i) Critical and creative thinking. (ii) Applying skills to issues. (iii) Action skills.

4. **Action orientation:** EE materials should promote civic responsibility, encouraging learners to use their knowledge, personal skills, and assessments of environmental issues as a basis for environmental problem solving and action. (i) Sense of personal stake and responsibility. (ii) Self-efficacy.

5. **Instructional soundness:** EE materials should rely on instructional techniques that create an effective learning environment. (i) Learner-centered instruction. (ii) Different ways of learning. (iii) Connection to learners' everyday lives. (iv) Expanded learning environment. (v) Interdisciplinary. (vi) Goals and objectives. (vii) Appropriateness for specific learning settings. (viii) Assessment.

6. **Usability:** EE materials should be well designed and easy to use. (i) Clarity and logic. (ii) Easy to use. (iii) Long lived. (iv) Adaptable. (v) Accompanied by instruction and support. (vi) Make substantiated claims. (vii) Fit with national, state, or local requirements.

Environmental education is not restricted to in-class lesson plans. There are numerous ways children can learn about the environment in which they live. From experiential lessons in the school yard and field trips to national parks to after-school green clubs and school wide sustainability projects, the environment is a topic which is readily and easily accessible. Furthermore, celebration of Earth Day or participation in EE week is a great way to dedicate your lessons to environmental education. To be most effective, promote a holistic approach and lead by example, using sustainable practices in the classroom and school grounds and encouraging students and parents to bring environmental education into their home.

The final aspect of environmental education policies, but certainly not least important, is training individuals to thrive in a sustainable society. In addition to building a strong relationship with nature, American citizens must have the skills and knowledge to succeed in a 21st century workforce. Thus, environmental education policies fund both teacher training and worker training initiatives. Teachers must be trained to effectively teach and incorporate environmental studies in their curricula. On the other hand, the current workforce must be trained or re-trained so that they can adapt to the new green economy. Environmental education policies that fund training programmes are critical in educating citizens to prosper in a sustainable society.

Related Disciplines

Environmental education has crossover with the disciplines of outdoor education and experiential education. Both disciplines complement environmental education yet have unique philosophies.

- Outdoor education means learning "in" and "for" the outdoors. It is a means of curriculum extension and enrichment through outdoor experiences." Environmental education is often taught or enhanced through outdoor experiences. The out of doors experience while not strictly environmental in nature often contain elements of teaching about the environment.
- Experiential education is a process through which a learner constructs knowledge, skill, and value from direct experiences" Experiential education can be viewed as both a process and method to deliver the ideas and skills associated with environmental education.

While each of these disciplines have their own objectives, there are points where both disciplines overlap with the intentions and philosophy of environmental education.

Trends

One of the current trends within environmental education seeks to move from an approach of ideology and activism to one that allows students to make informed decisions and take action based on experience as well as data. Within this process, environmental curricula have progressively been integrated into governmental education standards. Some environmental educators find this movement distressing and a move away from the original political and activist approach to environmental education while others find this approach more valid and accessible.

Movement

There is a movement that has progressed since the relatively recent founding of the idea of environmental education in industrial societies, which has transported the participant from nature appreciation and awareness to education for an ecologically sustainable future. This trend may be viewed as a microcosm of how many environmental education programmes seek to first engage with participants through developing a sense of nature appreciation which is then translated into actions that affect conservation and sustainability.

INSTITUTE OF TECHNOLOGY

Institute of technology, and polytechnic, are designations employed in a wide range of learning institutions awarding different types of degrees and operating often at variable levels of the educational system. It may be any institution of higher education and advanced research or vocational education, specializing in science, engineering and technology or different sorts of technical subjects. It may also refer to a secondary education school focused in vocational training. While the terms institute of technology and polytechnic are synonymous, the preference concerning which one is the preferred term varies from country to country.

Institutes of Technology *versus* Polytechnics

The institutes of technology and polytechnics have been existing at least since the 18th century, but became popular after World War II with the expansion of technical education, associated with the new needs created by industrialization. An exception is the *Ecole Polytechnique*, one of the oldests in the world, which has educated French *élites* since its foundation in 1794. In some cases, polytechnics or institutes of technology are engineering schools or technical colleges.

Sometimes, also institutes of technology are engineering and science research intense universities when they meet conditions necessary to be formally considered a university: autonomy to offer masters and doctoral degrees and independence as research institutions. In India, Indian Institutes of Technology are specific elite institutes which were based on a post WWII recommendation for industrialization. Those are highly regarded full chartered universities with a long history.

In several countries, like Germany, Switzerland and Turkey, institutes of technology and polytechnics are institutions of higher education, and have been accredited to award academic degrees and doctorates.

In countries like Iran, Finland, Malaysia, Portugal, Singapore or the United Kingdom, there is often a significant and confused distinction between polytechnics and universities. In the UK Polytechnics offered university equivalent degrees from Bachelors, Masters and PhD that were validated by the Council for National Academic awards.

In 1992 UK Polytechnics were designated as universities. In Ireland the term *institute of technology* is more favoured synonym of a *regional technical college* though the latter is the legally correct term; however, Dublin Institute of Technology is a university in all but name as it can confer degrees in accordance with law, Cork Institute of Technology and another of other Institutes of Technology have delegated authority from HETAC to make awards to and including Masters degree level.

In a number of countries, although being today generally considered similar institutions of higher learning across many countries, polytechnics and institutes of technology used to have a quite different statute among each other, its teaching competences and organizational history. In many cases *polytechnic* were a former designation for a vocational institution, before it has been granted the exclusive right to award academic degrees and can be truly called an *institute of technology*. A number of polytechnics providing higher education is simply a result of a formal upgrading from their original and historical role as intermediate technical education schools. In some situations, former polytechnics or other non-university institutions have emerged solely through an administrative change of statutes, which often included a name change with the introduction of new designations like *institute of technology, polytechnic university, university of applied sciences*, or *university of technology* for marketing purposes. Such emergence of so many upgraded polytechnics, former vocational education and technical schools converted into more university-like institutions has caused concern where the lack of specialized intermediate technical professionals lead to industrial skill shortages in some fields, being also associated to an increase of the graduate unemployment rate. This is mostly the case in those countries, where the education system is not controlled by the state and everybody can grant degrees. Evidence have also shown a decline in the general quality of teaching and graduate's preparation for the workplace, due to the fast-paced conversion of that technical institutions to more advanced higher level institutions.

Australia 1970s-1990s

During the 1970s to early 1990s, the term was used to describe state owned and funded technical schools that offered both vocational and higher education. They were part of the College of Advanced Education system. In the 1990s most of these merged with existing universities, or formed new ones of their own. These new universities often took the title University of Technology, for marketing rather than legal purposes. AVCC report The most prominent such university in each state founded the Australian Technology Network a few years later.

1990s-Today

Since the mid-1990s, the term has been applied to some technically minded Technical and Further Education (TAFE) institutes. These primarily offer vocational education, although some are beginning to offer higher education. This usage of the term is most prevalent in NSW and the ACT. The new terminology is apt given that this category of institution are becoming very much like the institutes of the 1970s-1990s period.

Belgium and the Netherlands

Hogeschool is used in Belgium and *Hogere Technische School* (HTS) in the Netherlands. The Hogeschool has many similarities to the *Fachhochschule* in the German language areas and to the *Ammattikorkeakoulu* in Finland.

Hogeschool institutions in the Flemish Community of Belgium (such as the Erasmus Hogeschool Brussel) are currently undergoing a process of academization. They form associations with a university and integrate research into the curriculum, which will allow them to deliver academic master's degrees.

In the Netherlands, four former institutes of technology have become universities over the past decades. These are the current three Technical Universities (at Delft, Eindhoven and Enschede), plus the former agricultural institute in Wageningen. A list of all *hogescholen* in the Netherlands, including some which might be called polytechnics, can be found here.

Brazil

In Brazil there is only one University of Technology: Federal University of Technology, Paraná (UTFPR) and the Institutes of Technology of the armed forces: Instituto Tecnológico de Aeronáutica and Instituto Militar de Engenharia.

Canada

- British Columbia Institute of Technology, (Burnaby, British Columbia)
- Kwantlen Polytechnic University, (Surrey, British Columbia)
- PAVI (Vancouver, British Columbia)
- Seneca College (Toronto, Ontario)
- Humber College (Toronto, Ontario)
- Southern Alberta Institute of Technology, SAIT (Calgary, Alberta) (Name changing to SAIT Polytechnic)
- Northern Alberta Institute of Technology, NAIT (Edmonton, Alberta)
- Saskatchewan Institute of Applied Science and Technology, SIAST (4 campuses: Regina, Saskatoon, Moose Jaw and Prince Albert)

- ETS (Montreal, Quebec)
- University of Ontario Institute of Technology (Oshawa, Ontario)

Finland — Ammattikorkeakoulu — Yrkeshogskola

An Ammattikorkeakoulu is the common term in Finland, as is the Swedish alternative "Yrkeshogskola" – their focus is on studies leading to bachelor degree, particularly in technology. After January 1 2006 some Finnish institutes of technology switched the English term *polytechnic* to the term *University of Applied Sciences* in their official names. The Ammattikorkeakoulu has many similarities to the *Hogeschool* in Belgium and in the Netherlands and to the *Fachhochschule* in the German language areas.

Some Finnish polytechnics are:

- Turku University of Applied Sciences
- Kemi-Tornio University of Applied Sciences

a complete list may be found in List of polytechnics in Finland

French Language Areas

French Language Areas — Instituts de technologie (Grandes écoles)

Collegiate universities grouping several engineering schools or multi-site clusters of French Grandes écoles provide sciences and technology curricula as autonomous higher education engineering institutes.

They include :

- Centrale Graduate School
- Grenoble Institute of Technology
- Institut national des sciences appliquées.
- Paris Institute of Technology

They provide science and technology master degrees and doctoral degrees.

French Language Areas — Universités technologiques/ Instituts universitaires de technologie / Polytech

France education system also includes three universities of technology:

- Université de technologie de Belfort-Montbéliard
- University of Technology of Compiègne
- University of Technology of Troyes

In addition, France education system includes many Institutes of technology, embedded within most French Universities. They are referred-to as *Institut Universitaire de technologie* (IUT). Instituts universitaires de technologie provide undergraduate technology curricula. *Polytech institutes*, embedded as a part of eleven French universities provide both undergraduate and graduate engineering curricula.

French Language Areas — Écoles Polytechniques

Higher education systems, that are influenced by the French education system set at the end of the 18th century, use a terminology derived by reference to the French École polytechnique.

German language areas - Technische Universitäten and Fachhochschulen

Technische Universität (abbreviation: *TU*) and *Fachhochschule* (abbreviation: *FH*) are the common terms for Universities of Technology or Universities of Applied Sciences, respectively, in a number of countries with German influences, these are Austria, Germany, Liechtenstein, Switzerland — the term *Fachhochschule* is now officially translated as "University of Applied Sciences" to respect their role. Polytechnic type institutions are widespread in this part of Europe. For example, Germany has 159 Fachhochschule-institutions.

There is a differentiation between a *Technische Universität* (or *Technische Hochschule*) and a *Fachhochschule*. A *Technische Universität* focuses more on research, has a much stronger emphasis on teaching theoretical background knowledge and can grant doctoral degrees as well as the *Diplom* degree, Bachelor's and Master's degrees. The largest and most renowned *Technische Universitäten* in Germany are members of TU9 German Institutes of Technology.

A *Fachhochschule,* in contrast, focuses more on teaching hands-on and has an emphasis on applications rather than theoretical background. Therefore, a *Fachhochschule* performs research in the field of applying science. It can grant Bachelor's, *Diplom (FH)* and Master's degrees. The *Diplom* degree of the Fachhochschule is officially a *Diplom (FH)* which is considered being less valuable than a university degree. However, through the Bologna process, the Bachelor's, Master's degrees have been made equivalent, although there still is a big difference in evaluation of these degrees concerning salary and reputation (alumni of *Technische Universitäten* are still evaluated as being more valuable).

Greece

In Greece there are 2 "Technical Universities", the National Technical University of Athens and the Technical University of Crete.

On the other hand, there are Greek technological educational institutes which do not have full university status. After the 2001 Higher Education Reform Act (N.1416) Technological Educational Institutes constitute a complementary part of the public higher education in Greece. They confer 4-year bachelor's degrees, but in practice their graduates have still fewer rights than universities'. Among other differences they have, the technological educational institutes are allowed to run graduate programmes only in collaboration with universities.

Hong Kong

The first polytechnic in Hong Kong is The Hong Kong Polytechnic, established in 1972 through upgrading the Hong Kong Technical College (Government Trade School before 1947). A second polytechnic, the City Polytechnic of Hong-Kong, was founded in 1984. These polytechnics awards diplomas, higher diplomas, as well as academic degrees. Like the United Kingdom, the two polytechnics were granted university status in 1994 and 1995 respectively, and renamed The Hong Kong Polytechnic University and the City University of Hong Kong. The Hong Kong University of Science and Technology, a university with a focus in applied science, engineering and business, was founded in 1991.

Hungary

One of the oldest (est. 1782) Universities of Technology of the world is located in Budapest.

- Budapest University of Technology and Economics
- Hungary's second University of Technology was Technical University of Heavy Industry in Miskolc (now it is called University of Miskolc). It was founded in Selmecbánya as Royal Hungarian Academy of Mining and Forestry in 1735.

India

Polytechnic is a technical institute which imparts technical education in India. Polytechnics are not affiliated to any university. They offer three year duration diploma courses in engineering. The courses offered in polytechnics can be said to be abridged version of degree courses offered in engineering colleges in India. The courses are designed in such a way that the students are able to perform basic engineering tasks. The diploma holders in engineering are generally employed as supervisors or junior engineers in the companies. The minimum qualification for admission to polytechnics is pass in SSLC (Standard Tenth). The polytechnics are affiliated to state technical boards. The All India Council of Technical Education is the regulating authority for polytechnics in India.

After successfully completing their diploma in polytechnic, students can gain lateral entry to engineering degree (under graduate) courses called BE which are conducted by engineering colleges affiliated to universities.

Polytechnic's in India

- Pusa Polytechnic
- MSPVL Polytechnic College
- Purulia Polytechnic, Purulia, West Bengal, India.

Indonesia

- Institut Teknologi Bandung (ITB)
- Institut Teknologi Telkom (IT Telkom)
- Politeknik Negeri Bandung (POLBAN)

Iran

- Tehran Polytechnic or Amirkabir University of Technology is a Full technology university in Tehran. Tehran Polytechnic
- Isfahan Institute of Technology is located in Isfahan.

Faculty of Technology and Engineering of Shiraz University. Faculty of Technology and Engineering of Tehran university. Faculty of Technology and Engineering of

Mashhad university. Faculty of Technology and Engineering of Tabriz university. Faculty of Technology and Engineering of Arak University. Sharif University of Technology and Engineering- Tehran. University of Sceince and Technology- Tehran. Khaje Nasir University of Science and Engineering. Sahand University of Technology— Tabriz.

Iraq

- University of Technology, Iraq
- Technical College, Basrah

Ireland

The Republic of Ireland has an "Institute of Technology" system, formerly referred to as Regional Technical College (RTCs) system — the latter term is still the correct legal term for the colleges when used generically or collectively. These institutions have a similar number of students attending as at Irish universities, and offer subdegree and degree level studies. Some institutions have "delegated authority" that allows them to make awards in their own name, after authorization by the Higher Education & Training Awards Council.

Dublin Institute of Technology developed separately from the Regional Technical College system, and after several decades of association with the University of Dublin, Trinity College it acquired the authority to confer its own degrees.

Israel

- Technion - Israel Institute of Technology

Italian Language Areas

In Italy the term *Politecnico* is used to refer to a university of applied sciences. Currently there are three *Politecnici* in the country:

- Politecnico di Milano

- Politecnico di Torino
- Politecnico di Bari

In the Italian-speaking part of Switzerland the term *Scuola Universitaria Professionnale* is used for the type of institution called *Fachhochschule* in the German-speaking part of the country.

Japan

See the Tokyo Institute of Technology and the Imperial College of Engineering, forerunner of The University of Tokyo's engineering faculty. See also Kyushu Institute of Technology, Osaka Institute of Technology and Nagoya Institute of Technology.

New Zealand

New Zealand polytechnics are established under the Education Act 1989 as amended, and are considered state-owned tertiary institutions along with universities, colleges of education, and wânanga; there is today often much crossover in courses and qualifications offered between all these types of Tertiary Education Institutions. Some have officially taken the title 'institute of technology' which is a term recognized in government strategies equal to that of the term 'polytechnic'. One has opted for the name 'Universal College of Learning' (UCOL), and another 'Unitec New Zealand'. These are legal names but not recognized terms like 'polytechnic' or 'institute of technology'. Many if not all now grant at least bachelor-level degrees.

Since the 1990s, there has been consolidation in New Zealand's state-owned tertiary education system. In the polytechnic sector: Wellington Polytechnic amalgamated with Massey University. The Central Institute of Technology explored a merger with the Waikato Institute of Technology, which was abandoned, but later, after financial concerns, controversially amalgamated with Hutt Valley Polytechnic, which in turn became Wellington Institute of Technology. Some smaller polytechnics in the North Island, such as Waiarapa Polytechnic, amalgamated with UCOL. (The only other amalgamations have been in the colleges of education.)

The Auckland University of Technology is the only polytechnic to have been elevated to university status; while Unitec has had repeated attempts blocked by government policy and consequent decisions; Unitec has not been able to convince the courts to overturn these decisions.

Pakistan

The Polytechnic institutes in Pakistan, offer a diploma spanning three years in different branches. Students are admitted to the diploma program based on their results in the 10th grade standardized exams. Main purpose of Polytechnic Institutes are to train technologists in various trades. These trades include Electrical, Mechanical, Civil, Auto and Diesel Technology, Electronics, Chemical, Refrigeration and Air Conditioning, Textile, Instrumentation etc.

These institutes are located throughout Pakistan and are in service since early 1950s.

Philippines

- Cebu Institute of Technology, a premier engineering school in the Philippines located in Cebu City, Philippines in the Visayas region. The University is known to have high selectivity in admissions as well as excellence in engineering research and education.
- Mapua Institute of Technology, an engineering school in the Philippines.
- Polytechnic University of the Philippines, a state university in the Philippines and is also referred to as

the National Comprehensive University of the Philippines.

- Technological University of the Philippines.
- Technological Institute of the Philippines.

Portugal

The designation "Institute of Technology" is not applied at all, being meaningless in Portugal. However, there are higher education educational institutions in Portugal since the 1980s, which are called polytechnics. After 1998, they were upgraded to institutions which are allowed to confer bachelor's degrees (the Portuguese *licenciatura*). Before then, they only awarded short-cycle degrees which were known as *bacharelatos* and didn't provide further education. After the Bologna Process in 2007, they have been allowed to offer 2nd cycle (masters's) degrees to its students. The polytechnical higher education system provides a more practical training and is profession-oriented, while the university higher education system has a strong theoretical basis and is highly research-oriented.

Russia

- Saint Petersburg Polytechnical University
- Tomsk Polytechnic University

Singapore

Singapore retains a system close to that applying in the United Kingdom from 1969–1992, distinguishing between polytechnics and universities, but also including a third component, the institute of technical education (ITE). Under this system, most Singaporean students sit for their 'O' Level examinations after a four or five years of education in secondary school, and apply for a place at either ITE, a polytechnic or a Pre-university centre (a junior college or the Millennia Institute, a centralized institute). A few secondary schools are now offering a six-year programme which leads directly to university entrance.

Polytechnics offer three year diploma courses in subjects such as information technology, engineering subjects and other vocational fields. There are a total of 5 polytechnics in Singapore. They are namely:

- Singapore Polytechnic
- Ngee Ann Polytechnic
- Temasek Polytechnic
- Nanyang Polytechnic
- Republic Polytechnic

Sri Lanka

- Sri Lanka Institute of Information Technology
- Institute of Technological Studies

South Africa

South Africa is in a process of transforming its "higher education landscape". Historically a division in South Africa between Universities and Technikons (polytechnics) as well between institutions servicing particular racial and language groupings. In 1993, Technikons were afforded the power to award certain technology degrees. Beginning in 2004 former Technikons have either been merged with traditional Universities to form Comprehensive Universities or have become Universities of Technology, however the Universities of Technology have not to date acquired all of the traditional rights and privileges of a University (such as the ability to confer a wide range of degrees).

Thailand

Most of Thailand's institutes of technology were developed from technical colleges, in the past could not grant bachelor's degrees; today, however, they are university level institutions, some of which can grant degrees to the doctoral level. Examples are Pathumwan Institute of Technology (developed from Pathumwan Technical School), King Mongkut's Institute of Technology Ladkrabang (Nondhaburi Telecommunications Training Centre), and King Mongkut's Institute of Technology North Bangkok (Thai-German Technical School).

Institutes of technology with different origins are Asian Institute of Technology, which developed from SEATO Graduate School of Engineering, and Sirindhorn International Institute of Technology, an engineering school of Thammasat University.

Turkey

- Istanbul Technical University, 1773, Istanbul
- Yýldýz Technical University, 1911, Istanbul
- Karadeniz Technical University, 1955, Trabzon
- Middle East Technical University, 1956, Ankara
- Ýzmir Institute of Technology, Ýzmir, 1992
- Gebze Institute of Technology, Kocaeli, 1992

United Kingdom

Polytechnics were tertiary education teaching institutions in England, Wales and Northern Ireland. Polytechnics offered university equivalent degrees (Bachelors, Masters, PhD) validated by the Council for National Academic Awards CNAA. The comparable institutions in Scotland were collectively referred to as Central Institutions. Britain's first Polytechnic (now the University of Westminster) was established in 1838 at Regent Street in London and its goal was to popularize engineering and science knowledge in Victorian Britain.

In 1956, some colleges of technology received the designation College of Advanced Technology. They became universities in the 1960s. The designation "Institute of Technology" was occasionally used, and there were two "Institutes of Science and Technology": UMIST and UWIST of the University of Wales. Loughborough University was called Loughborough University of Technology from 1966 to 1996, the only institution in the UK to have had such a designation.

United States

A handful of American universities include the phrases "Institute of Technology", "Polytechnic Institute", "Polytechnic University", or similar phrasing in their names; these are generally research-intensive universities with a focus on engineering, science and technology. Conversely, schools dubbed "technical colleges" or "technical institutes" generally provide post-secondary training in technical and mechanical fields focusing on training vocational skills primarily at a community college level—parallel and sometimes equivalent to the first two years at a bachelor's-granting institution.

Venezuela

Institutes of technology in Venezuela were developed in the 1950s as an option for post-Secondary education in technical and scientific courses, after the *polytechnic* French concepts. At that time, technical education was considered essential for the development of a sound middle class economy.

Nowadays, most of the *Institutos de Tecnología* are privately run businesses, with varying degrees of quality. They are widely regarded, sometimes incorrectly, as inferior to the university education.

Most of these institutes award diplomas after three or three and a half years of education. Few, if any *Institutos de Tecnología* have any research facilities.

●●

2
Strategies in Environmental Education

Today as we make plans for Rio + 10, we should remember that Rio itself was Stockholm + 20. But unlike Rio, which became the largest gathering of heads of State and Government ever to gather together, the position in 1972 was nowhere near the same. Mrs. Indira Gandhi, then Prime Minister of India was the only visiting head of Government at the Conference in Stockholm. This was not so much because India's deep commitment to the environment per say, but rather the awareness that unless development needs and those of poverty in particular, were dealt with simultaneously, there could be neither protection of the environment nor success in the programmes to achieve development.

Launching the World Conservation Strategy in India, Prime Minister Indira Gandhi reminded the audience that "the interest in conservation is the rediscovery of a truth well known to our sages. The Indian tradition teaches us that all forms of life: animal and plant are so closely liked that disturbance in one gives rise to imbalance in the other". The Indian Constitution laid down the responsibility of Government to protect and improve the environment and made it a "fundamental duty of every citizen to protect and improve the natural environment including forests, lakes, rivers and wildlife".

It was in this background that a Department of Environment was established by the Government of India in

1980 and a Ministry formed in 1985. The role of the Ministry was seen as a nodal agency in the Central Government for the "planning, promotion, co-ordination and overseeing the implementation of various environmental and forestry programmes". In order to achieve these tasks, the Ministry recognized the "creation of environmental awareness among all sectors of the country's population" . The Ministry in turn recognized environmental education as a key to the success of any overall environmental strategy and decided to help in the setting up of a 'Centre of Excellence' in EE to play the vital role of setting the pace and the agenda for EE in the country. The Centre for Environmental Education, CEE, was setup in 1984.

In this chapter we will discuss the principal strategies and their experiences in the context of India and CEE.

2. The Challenge of India

With over a billion people and at least 17 major languages, with poverty and low literacy levels, with over 650,000 primary schools, with a rapidly increasing population, the development and environmental challenge is enormous. But in spite of all the human pressures India continues to be rich in it's biodiversity. With over 127,000 species and a variety of habitats, India is one of the world's mega biodiversity centres. While several of the goals of development remain elusive, with better nutrition and health care, the life expectancy at birth which was a mere 24 years at the turn of the last century and 34 years at independence in 1947, today is close to 70 years. The agricultural revolution has ensured that the food grain production has not only kept pace with the growth in population but has grown faster. A factor of four while population has grown three times over the past fifty years. Industrial production has shown an even more rapid growth, about 15 times since independence. Per capita income has grown 45 times during this period.

India has made considerable strides in slowing down its population growth rate. But with all the efforts the current estimate is that our population will only stabilize somewhere the middle of the century, by which time we will have become the most populous country of the world. Till then we add a population of approximately one Australia to India every year. With about 16% of the world's population and a little over 2% of it's land, there is already enormous pressure on our resources. But while the population increase puts pressure on our resources, the pressure of "development" is perhaps even greater.

3. SWOT analysis for EE

Strengths

The Constitution and the Government's commitment to the environment along with the rich tradition of environmentally sound practices are an important backdrop under which the countries EE strategy has been evolved. The Central Government and every state within India, now has a Ministry of Environment. All education departments recognize EE as an essential part of education. The law courts have been sympathetic to environmental issues and the Supreme Court has passed a directive that all students must go through a compulsory course on the environment and the media must show free of cost a certain amount of programmes to create environmental awareness. India has a vast network of NGO's that are actively participating in the creation of awareness on development and environmental issues. Working on their own and with Governments they are the backbone of the strategy to create greater environmental awareness, especially that leading to environmental action.

Weaknesses

Illiteracy which is estimated at around 48% and a high dropout rate in the school system impede most educational programmes. There is also insufficient recognition of EE or for that matter education itself as an essential tool for development. While commercial communication in the form of advertising is recognized by the business community as essential, development communication and environmental education and communication still remains at a sub-critical level as far as funding is concerned. The slower visible impact of EE on improvements in the environment is an inherent "weakness"

Opportunities

The rapid growth of mass media and the coverage of radio and TV offer a major opportunity for EE. The increasing availability of the Internet is also an emerging opportunity. With the media giving more coverage to environmental issues and in particular highlighting environmental linkages with various calamities and degradation, people are more interested in learning about the environment.

Threats

With opportunity, the new and powerful media with their increasing dependence on advertising revenues, have sparked off a consumerism which will have its own impact on people and their behaviour. The advertising budgets are huge compared to the total resources being spent on EE. This "alternative" consumer messages can be seen as a threat to EE.

4. Evolving a Government Strategy

Roles of MoEF, MHRD and other Ministries

While the principal responsibility for Environmental Education through the formal educational system is with the Ministry of Human Resource Development, the parallel responsibility of EE through non-formal efforts lies with the Ministry of Environment. But these efforts are by no means exclusive. There is a large amount of co-operation between the Ministries and with the institutions associated with them.

EE in the Curriculum

The national system of education, as defined in the National Policy on Education, 1986, visualizes a national curricular framework which contains a common core including several elements having a direct bearing on tne natural and social environment of the pupils.

The National Policy on Education, 1986 (NPE) states that the "protection of the environment" is a value which along with certain other values, must from an integral part of the curriculum at all stages of education. The policy states: "There is a paramount need to create a consciousness of the environment. It must permeate all ages and all sections of society, beginning with the child. Environmental consciousness should inform teaching in schools and colleges. This aspect will be integrated in the entire educational process."

There has been special emphasis on the need to give importance to environmental education, and this has been kept in view while designing curricula, framing the syllabi and developing text books. The philosophy behind the curricula is that environmental education could be infused into the curricula especially at the primary level.

The National Council of Educational Training and Research (NCERT), the apex body in the area of developing curriculum for the formal education system. Currently there is a national debate going on about how to make EE not only widespread but also effective at all levels of education. The strategy under discussion envisages a four pronged approach. The four elements of the strategy are:

(i) Strengthening Infusion of EE

(ii) Teacher Training for effective EE

(iii) Introduction of Environment as a separate subject

(iv) Use of non-formal methods of EE through the involvement of NGOs

The success of the above strategy will depend upon the close synergy and partnership between the MHRD and the MOEF, their key institutions, State Governments, NGOs as well as educational institutions throughout the country. In addition, the four components are closely linked and one cannot be viewed in isolation from the other. The operationalization would need to keep this in view.

CEE is closely involved in this process, part of which is being supported by the World Bank. The first step which is ongoing is the analysis of curricula across India to review the degree and quality of infusion of EE. The exercise is being undertaken by a partner organization Bharati Vidyapeeth Institute of Environment Education and Research (BVIEER)

Environmental Orientation to School Education (EOSE)

EOSE, most EE programmes are designed and implemented with National and State level perspectives, hence the full intent to environmental education will not be achieved because environmental concerns and problems may be locale specific and, therefore do not include consideration of global solutions.

Therefore, the Ministry of Human Resource Development (MHRD) supports a centrally sponsored scheme under its financial assistance is provided to government and voluntary organizations and other institutions working in the field of environmental education to develop locale specific programmes and materials. Under this scheme, CEE is one of the designated nodal agencies. To date approximately projects to support development of locale specific programmes and materials by NGOs have been facilitated.

Centres of Excellence

One of the strategies of the Department was to establish "Centres of Excellence" for research, training and education which would play a pivotal role in strengthening the country's infrastructure and thereby its ability to deal with these issues. The Centre for Environment Education was setup in 1984 at Ahmedabad under this programme. A second Centre, the C.P. Ramaswamy Environmental Education Centre at Chennai was setup in 1988 under the same scheme. The main activities of both the Centres are the development of environmental resource materials, organization of training programmes and creation of environmental awareness among teachers, students and the general community.

National Environmental Awareness Programme (NEAC)

The Ministry of Environment and Forests (MOEF), Government of India started National Environment Awareness Campaign (NEAC) in 1986 with the aim of creating environmental awareness at all levels of society. Under NEAC, the Ministry provides financial assistance to selected non governmental organizations, education and training institutes, community organizations, etc. to create massive awareness among citizens of India. Diverse target groups ranging from students/youth/teachers to rural and tribal population, women, professionals and the general public are covered under this campaign. The Campaign programmes are basically composed of a spectrum of short duration programmes.

Environmental Information System (ENVIS)

The Ministry of Environment and Forests has setup ENVIS for different aspects of environment. The objective of this system are to provide national environmental information service to the users of originators, processors and disseminators; to build up storage, retrieval and dissemination capabilities; to promote support and assist education and personnel training programmes designed to enhance environmental information processing and utilizing capabilities; to promote exchange of information amongst developing countries. CEE has been designated as the National ENVIS Centre for Environmental Education.

Museums of Natural History

Setting up of Natural History of Museums as unique facilities to promote non-formal environmental education and to create environmental awareness among people is a strategy of the Ministry of Environment and Forests. The National Museum of Natural History is located at Delhi. Regional Museums are also being established at Mysore, Bhopal and Bhubaneshwar.

Eco-clubs in Schools

A non-formal pro-active system of imparting environmental education to school children by involving them in various environmental activities through the scheme of Eco-clubs has been evolved by the Ministry of Environment and Forests.

An Eco-club may be set up in a middle/high school and should consist of a minimum of 20 members and a maximum of 50 members, particularly interested in the conservation and protection of the environment, and willing to dedicate time and effort on a regular basis towards this end. The members may be drawn from students belonging to classes from VI to X. Each Eco-club will be in charge of an active teacher in the school concerned. The Ministry provides some finance support per annum per Eco-Club.

Promoting Mass Media

The Ministry provide financial assistance for the production of films, video-spots and other audio visuals on important environmental issues for creation of mass awareness.

These are normally telecasted by our National Television (Door Darshan) Channels. There are slots in our National TV Channel specially for environment.

5. The CEE Strategy

CEE is a national centre in the field of environmental education (EE). Its main objective is to promote, across a wide range of target audiences, environmental awareness—an understanding of natural processes ad the interdependence between human beings and environment—and concern leading to environmental action.

CEE's target audiences primarily include children, youth and the general community. Equally importantly it addresses its efforts to servicing the EE needs of special groups such as teachers and educators, journalists and communicators, business and industry, government departments and NGOs, planners and policy makers, in-service professionals and community-based organizations.

From its inception CEE's activities have been rooted in and guided by certain basic strategies for maximization of their effectiveness and impact.

Enabling Tools

Any attempt at developing uniform educational programmes and materials for a country as diverse as India is bound to run into difficulties. Therefore, adaptability to different geographic, cultural, social and economic contexts built into the basic design of CEE's programmes and materials. CEE's educational designing is somewhat like a saree, more a tool than a finished product, a tool that comes alive in the hands of each user, its application being the combined effort of the designer and the user.

"The saree is a designed piece of clothing worn all over India. Over the years very beautiful designs, patterns and textures have been printed and woven into the Indian Saree and yet, several thousand years of Indian history has not tried to stitch the saree. It is worn in many ways ad fits all sizes. It is equally good for working, dressing up or sleeping in. The final effort of the person who designs the cloth and the person who wears it – of the designer and the user. This is a very different concept from that of designing, say, a well-stitched dress. The garment either fits or doesn't fit and, where it fits, leaves little room for the wearer to be innovative in its use."

Multipliers

Strategies are designed with a built in multiplier effect to achieve maximum reach and impact. Strategies include building up of partnerships with key partners in government and voluntary sectors; using the mass media, seminars, workshops, etc. For e.g. CEE's communication strategy for media envisages optimal use of existing networks. The 20,000-odd daily newspapers and periodicals in the country constitute a powerful medium to reach a significant section of the urban population and opinion leaders in rural areas. CEE runs a network for environment-related news items, features and articles to newspapers, non-governmental organizations and individual subscribers. The material is supplied in English and Hindi, and translated and used by publications in many other languages.

The multiplier effect is considerable. Exhibitions form another effective medium for environmental communication. CEE develops and fabricates exhibits on diverse environment-related themes. Television and video facilities are becoming more widely available in the country CEE recognizes the effectiveness of this medium in environmental education.

The Agents of Change, Selecting Target Groups

While CEE's mandate is to reach out across a wide range of target audiences CEE has strategically selected from within these groups in order to achieve maximum reach and impact. For e.g. in order to reach a large number of school children, CEE works to capacity build teachers as primary agent for introducing and sustaining EE in schools. This is because a single teacher can reach out to at least 100 to 200 children every year over a period of 15-20 years of his/her teaching career. Another initial strategy was to target the middle schools (Std. V to VIII), age group 10-14 years as the entry point to introduce activity based learning by doing approaches for environmental education. This was keeping in mind that at this stage, while children are open to, and capable of grasping new ideas, they are also not too pressured by competitive examinations. Thus the school managements also are more willing to allow innovative approaches at this stage.

Partnerships with NGO's

The areas and target groups requiring environmental education are enormous and cannot be handled by one single institution. CEE has, therefore, developed a strategy of working with others. Partnerships utilizing complimentary strengths of other organizations to avoid duplication of effort and network effectively for synergistic convergence of ideas and goals.

One example of such a partnership is school cluster programmes. Developing a sustainable model of partnerships between CEE, government and non-government institutions to sustain and invigorate this network was a key strategy adopted from the start. To this end CEE developed a model of close partnerships between the local schools, local NGOs, State Departments of Education across the country. Today, under the school-cluster programme supported by MoEF year-round programmes goon in schools through 55 NGO-school clusters covering all States. Over 1000 schools are part of this network.

Not Reinventing the Wheel

CEE strongly feels that it is a waste of precious talent and resources to reinvent the wheel. CEE also strongly feels that blindly using materials developed elsewhere would defeat CEE's role as innovators. It would also possibly reduce the chances of successful and widespread use of the materials.

Successful adaptation involves a variety of inputs including a fair amount of imagination and creativity. It is putting the issue under discussion in the relevant environmental and educational context; finding ways in which the material can retain its excitement and validity, but still work within the constraints of the local educational system; it is finding ways in which the creative aspects of the material are not lost, but find expression in local forms. Some examples of such adaptation is Nature Scope India which is the adaptation of the American Nature Scope under an understanding with the National Wildlife Federation, USA. Another such collaboration is with the World Resources Institute, USA under which a series of manuals is being developed for college teachers to aid them in introducing environmental concepts, issues and perspectives to undergraduate students.

Field Demonstrations

CEE has through its field programmes designed especially for communities living in and around national parks and sanctuaries developed appropriate programmes aimed at upgradation around protected areas in order to reduce dependence on the protected areas and their conflict with the forest department. The programmes are also aimed at creating awareness among the communities about the ecological significance of the forest areas around which they live. For e.g. CEE has launched the first Eco-development programme around the Ranthambhore National Park in Rajasthan. A field office was setup in 1987 at Sawai Madhopur near Ranthambhore for implementation of various educational and developmental programmes in selected villages around the park. The focus of the programme has been on environmental improvement by people themselves through environmentally sound technologies. Among the areas for which communication/education programmes were developed to this end, were animal husbandry. Another project – Hingolgadh Ecodevelopment Project was initiated in six villages around the Hingolgadh nature Education Park in Rajkot district of Gujarat. Today the project continues activities in 15 such villages and has extended its activities to 25 other villages not directly dependent on the sanctuary.

Experiencing Nature

Sundarvan, a nature education park, is part of CEE. This facility seeks to expose urban children and adults to the beauty and harmony of nature and to awaken their love for flora and fauna. This is carried out through a number of educational programmes using small animals, nature camps and training camps. Regular snake shows are held in Sundarvan to dispel myths and educate visitors about the nature and behaviour of re0ptiles. Outreach programmes include taking small animals to schools for demonstrations.

Two permanent camp sites have been set up for nature education in the context of forest and marine environments respectively at Bakore and Beyt Dwarka in Gujarat. These are managed by Sundarvan. In total about 13000 participants have been exposed to nature programmes through these camps. Some special camps for special groups such as deaf and dumb children and under privileged children have also been offered. CEE Regional offices also conduct nature camps as part of their activities participants include students, NGOs and nature enthusiasts.

Sustainable Development

Some new initiatives were launched after the Rio Conference and in the light of Agenda 21 objectives of working towards sustainable development. Developing a comprehensive information system in key decision-making areas such as industrial technology, conservation of protected areas and urban development was reinforced. With a view to resolve the increasing conflict between development and environment goals in a wide variety of areas, CEE launched in 1996 a special initiative that targets policy makers in every field who make resource use decisions that have an impact upon the environment.

Centre for Environment Education seeks to raise the level of general awareness and provide policy support on key environment and development issues through its initiative Media for Sustainable Development (MSD). MSD projects and programmes identify and articulate issues of current policy relevance in the environment-development interface and develop them further through a process of wide public consultation and discussion.

A major book series funded by the Ford Foundation, the Swiss Development Cooperation and CEE on policy issues relating to sustainable development is being brought out by CEE. The books aim to educate and inform decision-makers and policy planners about a concern or issue related to ecologically sensitive and equitable development and to present alternative and innovative models of action for a variety of ecological situations. Work has also been initiated on an audio-visual series dealing with sustainable development alternatives for decision-makers in particular, and the community in general.

MSD also runs a monthly press feeder service CEE-NFS (CEE's News and Feature Service) which supplies well researched, ready to use articles, news items and features on a non-exclusive basis to over 600 newspapers and magazines across the country. CEE-NFS seeks to make optimal use of existing media networks for the improvement of environmental awareness. CEE-NFS also carries articles based on issues identified by the books under the Environment and Development series. As part of MSD's capacity building efforts, training in communicating environment and development concerns is provided through a specialized internship programme for graduate and post-graduates.

IT/ Web sites

Information Technology (IT) advances in the last few years has thrown up an entirely new medium, which is fast and economical. The Internet has also provided everybody an equal opportunity to put information on any topic irrespective of which part of the world one is geographically located in. The Internet has given us the World Wide Web, the E-mail and the Internet Relay Chat amongst many others. CEE decided to look at this as a potential medium of the future for EE.

CEE has already launched seven sites targeted at various groups. The sites and the groups they are targeted at are

http://www.ceeindia.org/ (about CEE)

http://www.cee.envirodebate.org/ (for decision-makers)

http://www.cee.greenteacher.org/ (for educators)

http://www.cee.kidsrgreen.org/ (for children)

http://www.cee.saseanee.org/ (the south and southeast asia network site)

http://www.cywen.com/ (for youth)

www.ceeindia.org/greenhousegases (about climate change)

Each of the sites uses a different style. Highlights of some of the sites are:

- The site for decision-makers has an online debate facility. The topic of the debate changes regularly and allows the visitor to put across their views on Environment and Development issues.
- The site for educators not only gives visitors an overview of EE, but also ahs tips and activity ideas on how to make EE more effective in the classroom.
- The site for children aims at not only providing information about our environment, but also provide opportunities for children to interact with one another and also with experts.
- The site for environmental educators in south and southeast asia provides an overview of EE in the region and who are the key people their areas. It also gives a list of professionals trained in EE through a Regional Capacity building programme of CEE, IUCN and WWF.

In addition to the Internet, CEE is using IT in its capacity building programmes. The Environmental Education Bank for example is a computerized database of activities, concepts, case studies and resources. The electronic storage of the vast illustrations resource built up over the years on CDs is another area. This has helped in reducing the risk involved in sending out original illustrations and also saving of time in retrieving data.

The future areas which CEE would like to look at is ways in which IT can be used for providing Computer Based Training on the web, as part of its capacity building activities. This becomes particularly important as the cost of getting participants is going to become increasingly expensive.

Networking all of CEEs offices to enable faster and reliable systems of management, thus providing all offices to participate in discussions and meetings without having to get everybody to one location.

IT has many advantages, but care is to be taken so that it does not replace interpersonal communication avenues completely.

Training and Capacity Building

Over the years, training to capacity build individuals and institutions to environmental education and communication has become an important thrust area of CEE. A range of short and long term training programmes are offered. Some of the programmes are: Training programme in Environmental Education (TEE), Internship Programme in Environmental Journalism, Training in Librarianship and Documentation (TLD), Training for Indian Forest Officers, Environmental Education Bank, Certificate Course in Environmental Education (CCEE) to fulfill the requirement of trained professionals in EE in South And South East Asia.

From its inception, all these efforts and strategies have been part of CEE's philosophy of networking. Networking is not only a key activity but is characteristic of all projects and programmes at CEE. It is seen as a way to gain access to a greater range of expertise and other resources, and avoid the duplication of efforts thus leading to the efficient use of available resources.

Regional and International Approach

Various international activities are carried out by the Cenıre. The activities include on-going training programmes, workshops in collaboration with other NGOs, Institutes etc. The activities/programmes include Workshops/Seminars/ Specific Focus Meetings, Training Programmes, Subject

Expertise, Collaborative Projects and Partnerships and Networking.

SASEANEE is the Asian Regional Network of IUCN's Commission on Education and Communication. SASEANEE is a network of agencies and individuals in or interested in networking, initiating, or supporting environmental education programmes in the Region. The Secretariat is located at CEE. The South Asia Cooperative Environment Programme (SACEP), has through the Ministry of Environment and Forests, designated CEE as it's Subject Matter Focal Point of Environmental Education and Training and National Focal Point.

A number of workshops have been conducted in collaboration with International Agencies, Institutions, NGOs. Resource materials brought out by CEE have been adapted and developed for use in neighbouring countries by agencies of those countries. CEE has entered into various collaborative projects with international partners whereby there is sharing of resource materials and programmes. Such collaborative projects help in sharing and enriching the perspectives of all the partners concerned. These include partnerships with the following: National Wildlife Federation, USA, State University of New York, US National Park Service, World Resources Institute, USA, Field Studies Council, UK, International Television Trust for the Environment (TVE), UK and UNESCO-UNEP.

6. Conclusion

Speaking in 1928, Mahatma Gandhi cautioned that "God forbid that India should even take to industrialism after the manner of the west.... It would strip the world bare like locusts." And yet development is a must. P.V. Narasimha Rao, India's Prime Minister at the time of Rio emphasized the need of country to "take upon itself the challenge to deal with population-related issues which are inextricably linked to the total development of our nation." But with all these warnings, India like most other developing countries is seeking to find sustainable strategies to an unsustainable goal. India's Report to the United Nations' Conference on Environment and Development. 1992 highlighted this dilemma that "While the challenge of development might seem to be 'how to get there', the real challenge is 'how not to get there'.

With increasing globalization, rapidly increasing consumption levels fueled by big budget advertising and a rate of population growth that is yet to come down to sustainable levels, the challenge of EE in India are enormous. Several strategies are in place and new ones are being evolved. But whether these will be adequate and are able to act in time is a question that only time can tell.

3

Challenges in India

India, a country with the seventh largest landmass in the world, is a land of ancient traditions.With over a billion people and at least 17 major languages, the diversity of India in terms of culture and biological wealth is enormous. In spite of rapidly changing lifestyles, the traditions of living in harmony with nature and of environmentally sound practices underpin the lives of most people. It is against this backdrop that the country's EE strategy has been evolved. The Constitution of India explicitly makes environmental conservation a duty. The Central Government and all states within India now have a Ministry or Department of Environment. Education departments recognize EE as an essential part of education. The law courts of the country have been sympathetic to environmental causes. India has a very large number of very active NGOs who are involved in a variety of activities from policy analysis to school programmes; from participatory natural resources management to activism.

India continues to be rich in its biodiversity. Agricultural revolution has ensured that food-grain production has not just kept pace with the population increase, but has rather grown faster. With better nutrition and health care, the life expectancy has gone up by almost 3 times since the independence. The per capita income has grown almost 45 times during this period.

But the environment in India faces several challenges too. With poverty and low literacy levels, over 650,000 primary schools and rapidly increasing population, the development and environmental challenges are enormous. India has made considerable strides in slowing down its population growth. But with all efforts, India's annual population increase is equal to the population of Australia. With about 16 per cent of the world population and a little over 2 per cent of its land, there is already enormous pressure on our resources. But while the population increase puts pressure on resources, the pressure of 'development' is perhaps even greater.

The Role of Environmental Education in the Indian Context

Thus there are many challenges for environmental educators in India. Apart from the obvious ones of helping strengthen environmental management and conservation, one of the important ones is to bring about awareness of the need that the country develops in less wasteful ways than is the current paradigm. EE is one of the tools that can help India achieve this goal. There are considerable initiatives in EE in India today. There are also several challenges. Some of these are:

- The challenge, in a large and diverse country, to find the right blend between centralized and de-centralized efforts and approaches.
- The challenge of reaching out to large numbers cost-effectively The challenge of making environmental considerations relevant and meaningful to various groups.
- The challenge of putting EE on the agenda of educational decision-makers.
- The challenge of putting sustainable development concerns high on the agenda of policy makers, and
- Finding and developing human and financial resources for EE.

This Chapter, through a few case examples, attempts to reflect the kind of environmental challenges India faces, and the wide and rich variety of EE efforts that have been initiated in the country, and also tries to analyze what has worked and why, or for that matter even what has not worked and why.

Appreciation and concern for the environment are values that need to be inculcated during the early years of development. And thus EE for children and youth becomes an integral and important part of the EE strategy of any country. India, a land of diversity ecological, biogeographical, cultural, social requires efforts which are rooted in the local context, which present a picture to which the child relates, can provide solutions of which the child can be a part. The first case of this paper highlights one such pioneering effort.

1. Tailoring Education to the Context

Almora: Where Nature Bestows its Beauty

Uttarakhand, the 27th state of the Republic of India, is blessed with magnificent glaciers and majestic rivers. In this state of mountains and rivers is situated the Uttarakhand region comprising eight hill districts of the state Uttarkashi, Dehradun, Tehri, Chamoli, Pauri, Almora, Nainital and Pithorgarh.

Famous for its rich cultural heritage, fairs and festivals, and magnificent wildlife, Uttarakhand constitutes a unique and fragile ecosystem. The region represents a transition between the drier west Himalayas and the moist east Himalayas, and is home to unique flora and faunathe Himalayan Tahr, Himalayan Mouse Hare, Musk Deer, Snow Cock, Civet Cat and a number of pheasants, like the Monal Pheasant, Chir Pheasant, and White crested Khalij. The mighty Ganges and several other important rivers of India originate here. The area, with few developmental pressures, has been declared as the Uttarakhand Biosphere Reserve. The exotic Valley of Flowers (a National Park) is part of the Biosphere Reserve.

Standing at a height of 1638m, in this picturesque region of Uttarakhand, is the district of Almora the district for which once Gandhiji said,

> "In these hills, nature's hospitality eclipses all men can ever do. The enchanting beauty of Himalayas, their bracing climate and soothing green that envelopes you leaves nothing more to be desired. I wonder whether the scenery of these hills and the climate are to be surpassed, if equaled by any beauty spots of the world. After having been nearly three weeks in Almora Hills, I am more than ever amazed why our people need go to Europe in search of health."

Can Almora Take It?

While not too many activities have yet happened in the region, population pressures (human and cattle) and changing lifestyles pose a threat to the delicate balance of this sub-alpine forest ecosystem. The results of overgrazing and the slow but real change of lifestyles are slowly becoming visible. Several hillocks have become barren, with no trees or greenery left. Landslides too have become more frequent. Human-animal conflicts are also being reported.

Our Land, Our Life: The Genesis of a New Idea

The National Policy on Education (1986) of India states that protection of environment is a value which, along with certain other values, must form an integral part of the curriculum at all stages of education. India, like several other countries has taken the route of infusion of concepts and information about the environment into existing subjects/ courses beyond the primary school level. It is recognized that the creation of awareness is only the first step in what must be a much more comprehensive EE process. But it is not often possible, through a school environmental programme to bring in linkages of the environment with economic and political determinants of the problems, with people's livelihood, or to develop practical skills that will enable students to effectively tackle these problems.

Another challenge is how to make EE effective and locale-specific. Generally within a state, one set of syllabi and textbooks are used, though environmental conditions and environmental concerns vary from one region of the state to the other. Since environmental concerns are locale-specific, and therefore do not admit of global solutions, it is necessary to supplement the efforts at national and state levels with a more intensive locale-specific effort.

'Our Land, Our Life' is an experiment for developing and testing a separate course of EE that focuses on the local livelihood aspects of environmental problems, and on developing practical skills for solving them. The experiment is being carried out in the state of Uttarakhand by the Uttarakhand Seva Nidhi Environmental Education Centre (USNEEC), Almora, with support from the Ministry of Human Resource Development, Government of India and the Department of Education, Uttarakhand.

The course, originally designed in 1987, and subsequently revised several times in the light of experience, is currently running in 550 government schools in the state in Classes 6-8 (age groups 12-14). In the year 2001, the Uttarankhand Government has taken a decision to introduce this course in all the schools and intermediate colleges in the state, and gradually to extend it to Classes 9 to 12.

This Course addresses land degradation (deforestation and soil erosion), the chief environmental problem of the state, as well as other problems arising out of village land degradation, i.e. water scarcity, falling crop yields, fuelwood and fodder scarcity, drudgery for women and girls, and the forced migration of men and boys for employment. The course meets the objective of local relevance and focuses on immediate issues of well-being and livelihoods. The local community is involved in the course. Skill development is emphasized through practical work. Whereas an infusion approach seeks to give an environmental orientation to the subject matter of conventional school disciplines, the course approach, as formulated by the USNEEC, draws concepts and knowledge from these disciplines, and brings them to bear on the solution of local environmental/livelihood problems.

The Course Design

The course is designed for Classes 6 to 8. It consists of 36 practical exercises in all. The broad subject-matter themes of these are: land, soil, water, trees, crops, compost, animals, fodder, fuelwood, people and the village ecosystem of which all these form parts of. The exercises are set forth in three workbooks. The objective of each exercise is stated and the methodology to be employed is explained. Spaces are provided for answering questions, doing calculations, making diagrams, maps and notes. Most exercises are accompanied by theory lessons that explain the concepts and procedures needed to do the exercises meaningfully. Students generate their own data for the most part, but some supplementary data is also given in various theory lessons where there is a need to make a concept clear or make it possible for the students to complete an exercise.

The set of three workbooks is accompanied by a Teacher's Manual. This sets forth the rationale of the course, its objectives, the concepts to be highlighted each year, and suggestions for carrying out the various exercises.

The Course is designed to require five short periods (40 minutes each) per week and one long period (3 hours per month).About half the long periods are spent in nearby study villages making measurements and observations, and talking to residents. The unique features of the course are briefly described below.

Locale specificity and relevance: The Course is designed so that it is practical in nature, locale-specific, and promotes a holistic understanding of the Uttarakhand village environment, as Uttarakhand is primarily rural. The strength of the program is the linkages it establishes between the daily life and activities of the students, and their environment.

Holistic approach: The students are encouraged to treat the village as a single ecosystem and are taught how to manage this system for maximum productivity. The course content is designed and treated in a manner so as to communicate the concept of ecosystem and other derived concepts like ecosystem evolution, equilibrium, degradation, sustainability, carrying capacity and a number of other critical ecological concepts. But all this is communicated without losing the focus on the village ecosystem. Larger ecosystems are then explained in connection to and in relation to the local village system.

Awareness and action: Another strength of the course is its focus on both awareness and action. The course ultimately aims to prepare the children to be effective future managers of the rural environment. The Course has a major component of practical work. This ensures that knowledge gained is applied and practiced. The EE Course culminates in comprehensive a forestation and irrigation plans made by the students, which can be implemented by the residents of the study village.

Support for teachers: Sustainability of the programme depends on its acceptance by teachers and their ability to deliver the course effectively. Teacher training is thus a priority within the initiative. Five-day camps are held from time to time for training teachers. There is also a manual for teachers. To enable them to guide students in the practical work, self-explanatory and self-contained workbooks are also provided.

Major Outcomes and Lesson Learned

- The course was initiated on a trial basis in about 40 selected schools. Today, over 500 schools are a part of the programme.
- The State Department of Education has made the course compulsory for all government schools in the state.
- The course has been allotted a separate period and is being evaluated too. The status of a separate subject helps the course being taken seriously by teachers and students and ensures a sharp focus on relevant issues.
- Over the years, UKSN activities have grown from an exclusive school-based initiative to other relevant community based programmes for women and children.With this, associated changes are also taking place, like women taking charge of community property resource management.

A major lesson learned is that to design an effective course for any particular region, the main problems of that region must be identified and understood. The study of national and global problems alone will not be meaningful to students. It is important to focus on the problems that are obvious, those that are jeopardizing the future well-being of students, because these are the problems about which they feel they need to do something.

Scope for Replication

The 'Our Land, Our Life' experiment is a successful example of making EE more effective and meaningful to the learner by making it locale-specific. The experience has much to offer for other areas with fragile environments.

The need for environmental education has been explicitly recognized by government policies of several countries. In many cases, policy directives exist for the incorporation of environmental education and communication (EE&C) into mainstream education. While governments are the policy making agencies, NGOs and CBOs are often the source of innovations. They are close to and responsive to community needs and can effectively implement programmes. Thus Government systems and the NGO networks are complementary to each other and a partnership between the two can make EE more effective.

2. Strategic Partnership for Institution Building

Protection of the Environment: A Tradition in India

Concern for nature and natural resources is not new for India. Respecting nature and living in harmony with it have long been parts of the Indian civilization. Launching the World Conservation Strategy in India, Prime Minister Indira Gandhi reminded the audience that *"the interest in conservation is the rediscovery of a truth well known to our sages. The Indian tradition teaches us that all forms of life: animal and plant are so closely linked that disturbance in one gives rise to imbalance in the other"* (Indira Gandhi, World Conservation Strategy for India, March 1988). The Indian tradition emphasizes living in harmony with nature. The Bishnois, for example, followers of a Rajput saint, Jambeshwar Maharaj, who lived towards the end of the fifteenth century, emphasize vegetarianism, non-violence, protection of trees and respect for all living things. In 1730, 363 Bishnois of Khejadli village, mostly women and old men, laid down their lives in an effort to protect trees being cut on the orders of the King of Jodhpur. Sacred groves are a unique tradition that has been responsible for islands of biodiversity in various parts of the country. Ashoka's pillar edict, dating back to 272-232 BC, proclaims protection for plants and animals.

State Support for the Environment

The Indian Constitution laid down the responsibility of the Government to protect and improve the environment (Constitution of India, Article 48-A) and made it a *"fundamental duty of every citizen to protect and improve the natural environment including forests, lakes, rivers and wildlife"* (Constitution of India, Article 51-G).

Mrs. Indira Gandhi, the then Prime Minister of India, was the only visiting head of State to attend the Conference on Human Environment at Stockholm, 1972. It is on this occasion that she first brought to fore the connection between environment and poverty. It was following this Conference that a Department of Environment was established by the Government of India in 1980, to act as the nodal agency for planning, promotion and coordination of various environmental programmes in the country. Since its inception, the Department has emphasized the promotion of environmental education at different levels, making it a people's movement. In 1983-84, the Department launched a new scheme on environmental education and awareness under which several activities, like seminars, group discussions, refresher courses, multimedia campaigns, eco-development camps etc. to various target groups, such as decision makers, politicians, administrators, teachers and students and general public were undertaken.

A National Institution is Created

One of the strengths of India's environmental movements is the vast network of NGOs that it has. Many of these NGOs have been actively participating in the creation of awareness on development and environmental issues. NGOs have brought about many innovations in EE and have been able to reach out effectively to various target groups. The Department, when it was created, recognized that it would have to build a number of partnerships to achieve the huge tasks ahead. New ways of working were being defined—there was openness to experiment.

It was in this spirit that, when the Department, after having seen the exciting work in environmental education being done by the Nehru Foundation for Development (NFD) thought of building a partnership with NFD. NFD was, through its three constituent activities and institutions VIKSAT (Vikram Sarabhai Center for Development Interaction), VASCSC (Dr. Vikram A. Sarabhai Community Science Center), and CHETNA (Center for Health Education, Training and Nutrition Awareness), involved in environment, science and health education for a variety of target groups, using a variety of media.

'What do you need to do this type and this quality of EE work at the national level?' was the question thrown to the core group at NFD by the then Secretary of the Department of the Government of India.

'Adequate resources, and freedom to work' was the answer.

Long discussions on the vision, mission and mandate of a national center were then discussed between the Government and NFD. A detailed proposal and a plan were developed and subsequently sanctioned. In 1984, a national centre for EE, the Centre for Environment Education (CEE) was established as a unique partnership between government and NGO, to play a pivotal role in setting the pace and agenda for EE in the country. Built around the core programmes and people of NFD, CEE was the first institution to be established under the Centers of Excellence programmes of the Department of Environment.

Strategies to Maximize Effectiveness and Impact

Recognizing that to fulfil its mandate as a national institution, it would have to develop EE programme to address a wide variety of sectors, the thrust areas of CEE's programme include (1) EE through schools, (2) EE through higher education, (3) EE through mass media, (4) EE through experiencing nature, (5) EE through interpretation, (6) EE for environment and development, (7) EE for industry, (8) EE for natural resource management, (8) EE in the urban context, (9) training, and (10) networking.

Given the challenges of India's diversity and numbers, CEE's activities and initiatives have been rooted in and guided by certain basic strategies so as to enhance the reach and impact of its programmes. These include:

People are key: CEE is built on the belief that ultimately, an organization is as good as the people it is made up of, and that people will give their best when they have the freedom to work and dream.

Adaptability: Given the geographical, cultural, social and economic diversity in India, flexibility is an essential basic in the design of programmes and materials. Adequate flexibility is built into CEE's educational programmes and materials to allow necessary adaptations.

Partnerships: Recognizing that to be effective in a country of the size and diversity of India partnerships are essential, such relationships to utilize complementary strengths of other organizations are built to avoid duplication of effort and to achieve synergy. The importance of involving concerned GOs and NGOs in strategic partnerships is recognized. One example of such a partnership is the school cluster programmes. To this end CEE developed a model of close partnerships between the local schools, local NGOs, State Departments of Education across the country.

Multiplier effect: A multiplier effect is built into the design of strategies, in order to achieve maximum reach and impact. For example, in order to capitalize on the reach of mainstream newspapers and magazines, CEE runs environmental news and features service which feeds into the mainstream papers.

Making use of existing opportunities: The stress is on using opportunities for education and communication wherever they exist. For example, CEE's interpretive programmes at natural and cultural heritage sites and at facilities take advantage of the fact that millions of people visit natural and cultural heritage sites.

Use of media and technology: CEE's programmes aim to use appropriate technology and media to meet the objectives of EE in any particular situation. The importance of using latest technological developments to leapfrog and achieve a wide reach is recognized.

Facilitating networking: In order to facilitate networking, CEE brings out a number of tools including directories, newsletters, bibliographies etc.

Not re-inventing the wheel: CEE tries to ensure that its programmes do not start from scratch, but rather, that they build on the experiences and learning of other groups. For example, Nature Scope India is an adaptation of the American teacher magazine, Nature Scope, brought out under an understanding with the National Wildlife Federation, USA.

Regional presence: CEE tries to have a presence at key locations, to ensure the reach of the Centre's activities to all parts of the country and beyond. CEE, over the years, has set up several regional cells and field offices.

International experience: Undertaking international collaborations is seen as a way of enhancing the quality, depth and range of programmes. Such collaborative efforts include partnerships with agencies, like National Wildlife Federation, USA, State University of New York, US National Park Service, World Resources Institute, USA, Field Studies Council, UK, International Television Trust for the Environment (TVE), UK and UNESCO-UNEP. CEE is one of the subject matter focal points on EE and training of the South Asia Co operative Environment Program (SACEP).

Identifying key entry points: While CEE's mandate is to reach out across a wide range of target audiences, CEE has strategically selected from within these groups in order to achieve maximum reach and impact. In the case of school programmes, it was decided to concentrate on the middle school level, since the barriers to entry to this level are the least, while the possible impact of intervention is very high.

Development of EE professionals: Capacity building in order to develop a cadre of professionals to improve the quality of EE, and its contribution to environmental conservation and sustainable development, is seen as a key activity. CEE conducts a variety of short and long duration training programmes for in service professionals as well as for students interested in pursuing a career in EE. Some examples of CEE's training programmes include the training programme in environmental education (TEE), internship programme in environmental journalism, training in librarianship and documentation (TLD), training for Indian forest officers, and an international certificate course in environmental education (CCEE) to fulfil the requirement of trained professionals in EE in South and Southeast Asia.

Non-exclusive organizational structure: The organization structure encourages initiative, autonomy, and inter-disciplinary approaches. The conscious effort is to develop different ways of doing things, according to interests, opportunities and expertise.

In-house infra-structural support: CEE has built up expertise/ facilities, e.g., studios, workshops, printing, editorial services, design services etc., to ensure innovation, research and development (R&D), quality control and excellence.

The Governance of CEE

Reflecting its unique structure, CEE's Governing Council has eminent persons in the field of education, environment, communication and management, as well as representatives of the Ministry of Environment and Forests, Government of India, and of the Department of Environment and Forests, Government of Gujarat.

The plethora of eminent professionals from a wide variety of field gives the Council, which is appointed by the Nehru Foundation for Development, is a unique strength. The Council guides programmes, approves budget and designs policy.

CEE is headed by the Director, who is the chief executive of a working team consisting of three sections; programme, technical and administrative staff.

Learning and Achievements

An institution that was set-up by the government with a vision that it would play a lead role in generating awareness about nature and nature conservation and would act as a model institution in EE and communication, has demonstrated that building partnerships is an effective strategy for sustaining EE and communication efforts.

CEE has created many model programmes and developed methodologies that are widely used in this country and abroad. It has developed educational strategies that are uniquely suited to a diverse country such as India. It has developed tools and training programmes that build capacity and empower many other groups to undertake EE work. Educational materials developed by CEE are being used and adapted by various groups in India as well as other countries. It has built partnerships that are essential for the enormous task before it.

CEE: Some Achievements

India's first Interpretation Programme at Kanha National Park, Madhya Pradesh, India was designed, developed and set up by CEE. CEE has developed and brought out over a hundred publications, 36 audiovisual materials and several exhibitions, directories and bibliographies in environmental educational and communication. These are in several Indian languages. Some of these have been adapted for use in other countries as well. CEE, as a part of its regional capacity program, has trained over 120 environmental education professionals from 22 countries. It is the subject matter focal point for environmental education and training of the South Asia Co-operative Environment Program. CEE was awarded the 'Indira Gandhi Paryavaran Puraskar' (IGPP) for the year 1997 by the Government of India, Ministry of Environment and Forests, for its significant contribution in the field of environmental education.

Some of the key lessons include : (1) GO-NGOpartnerships have the potential for building on the strengths of each sector. Such a partnership has to be based on mutual respect and the two partners have to interact as equals. (2) The relationship needs constant management. (3) Institution building is an ongoing activity and not one-shot effort. (4) Professionals given space, freedom and a challenging environment will give their best.

Scope for Replication

Building partnerships and complementing each other's efforts can maximize output, making efficient use of available resources. Efforts like this one can play a significant role in EE in a country. In India too, several centers of excellence including a second one on EE have been established. It is often assumed that economic, scientific and technological development and environmental concerns cannot go together. However, environment and development need not be seen as contradictions. This is the basis of the concept of sustainable development. EE is one of the tools to help achieve sustainable development. The next example illustrates how EE can help people make informed decisions. An effective EE campaign, such as the Silent Valley Movement can not only help people understand the need for harmony between environment and development but also convince the decision making authorities that alternatives exist, alternatives that have good economic and social outputs, with less impact on the environment. The Silent Valley Campaign is a landmark in India's environmental movement.

4

Ecology and Science of Earth

Ecology is the interdisciplinary scientific study of life processes explaining adaptations, external relations and interaction among organisms, the flux of materials and energy through living communities, the successional development of ecosystems, and the abundance and distribution of biodiversity in context of the environment.

Ecology is distinct from natural history because there are emergent phenomena operating at different environmental scales of study, ranging from molecular to galactic spheres of influence. An understanding of these phenomena require a conceptual distinction between ecology and the environment. Ecology is often misused as a synonym for environment, but it is one of the few academic disciplines dedicated to holism. In contrast, the environment describes all factors and scales of study that are external to the organism, including abiotic factors such as temperature, radiation, light, chemistry, climate and geology, and biotic factors, including genes, cells, organisms, members of the same species (conspecifics) and other species that share a habitat. Ecosystems are generally studied at the local community scale, such as measurements of primary producation in a wetland or the analysis of predator-prey dynamics affecting amphibian biomass. Ecological relations also regulate the flux of energy, nutrients, and climate all the way up to the planetary scale. The global sum of ecosystems is known as the biosphere where ecological theory has been used to explain self emergent regulatory phenomena at the planetary scale. This is known as the Gaia hypothesis. The Gaia hypothesis is an example of holism in ecology because it tests for principals relating to an evolving and self regulating planetary ecosystem that requires different explanations than those governing ecosystems at a smaller scale, such as a single wetland.

Examples of objects of ecological study include: Population processes, including reproductive behavior, mortality, bioenergetics and migrations, biomagnification, interspecific interactions such as predation, competition, parasitism and mutualism, plant and animal community structures and their function and resilience, and biogeochemical cycling. Because of its vast scope, ecological science is often closely related to other disciplines. Thus, molecular ecology addresses ecological questions using tools from genetics, paleoecology uses tools from archeology, and theoretical ecologists use often highly complex mathematical models to explore how ecosystems and their elements function.

Aside from pure scientific inquiry, ecology is also a highly applied science. Much of natural resource management, such as forestry, fisheries, wildlife management and habitat conservation is directly related to ecological sciences and many problems in agriculture, urban development and public health are informed by ecological considerations.

The term "ecology" has also been appropriated for philosophical ideologies like social ecology and deep ecology and is sometimes used as a synonym for the natural environment or environmentalism. Likewise "ecological" is often taken in the sense of environmentally friendly.

Historical Roots of Ecology

Ecology as a scientific discipline is relatively young, reaching prominence mostly in the second half of the 20th century. However, systematic ecological studies can trace roots to ancient times, with Aristotle and Theophrastus, for example, making early observations on animal migrations and plant biogeography respectively. Several notable 19th century scientists such as Alexander Humboldt (1769–1859), Charles Darwin (1809–1882), Alfred Russel Wallace (1823–1913) and Karl Möbius (1825–1908) made many important contributions, from laying down the foundation of biogeography to identifying an interacting groups of organisms as a functionally connected community (biocoenosis).

The term "ecology" itself (German: *Oekologie*) was first coined by the German biologist Ernst Haeckel in 1866, who defined it as "the comprehensive science of the relationship of the organism to the environment." The first significant textbook on the subject (together with the first university course) was written by the Danish botanist, Eugenius Warming. For this early work, Warming is sometimes identified as the founder of ecology.

Scope

Ecology is usually considered as a branch of biology, the general science that studies living organisms. It is associated with the highest levels of biological organization, including the individual organism, the population, the ecological community, the ecosystem and the biosphere as a whole. When referring to the study of a single species, a distinction is often made between its "ecology" and its "biology". For example, "polar bear biology" might include the study of the polar bear's physiology, morphology, pathology and ontogeny, whereas "polar bear ecology" would include a study of its prey species, its population and metapopulation status, distribution, dependence on environmental conditions, etc.

Because of its focus on the interrelations between organisms and their environment, ecology is a multidisciplinary science that draws on many other branches, including geology and geography, meteorology, soil science, genetics, chemistry, physics, mathematics and statistics. Due to its breadth of scope, ecology is considered by some to be a holistic science, one that over-arches older disciplines such as biology which in this view become sub-disciplines contributing to ecological knowledge. It has been argued that the mechanistic models which have driven the development of most other sciences are inappropriate for unraveling the complex interactions in most ecosystems, and that progress in ecology is better served by a central paradigm driven by information theory and complexity theory.

Ecology is also a highly applied science, especially with respect to issues of natural resource management. Efforts related to wildlife conservation, habitat management, mitigation of ecological impacts of environmental pollution, ecosystem restoration, species reintroductions, fisheries, forestry and game management are often the direct domain of applied ecology. Urban development, agricultural and public health issues are also often informed by ecological perspectives and analysis.

Disciplines

Ecology is a broad discipline comprising many sub-disciplines. A common, broad classification, moving from lowest to highest complexity, where complexity is defined as the number of entities and processes in the system under study, is:

- Ecophysiology examines how the physiological functions of organisms influence the way they interact with the environment, both biotic and abiotic.
- Behavioral ecology examines the roles of behaviour in enabling an animal to adapt to its environment.
- Population ecology studies the dynamics of populations of a single species.
- Community ecology (or synecology) focuses on the interactions between species within an ecological community.

- Ecosystem ecology studies the flows of energy and matter through the biotic and abiotic components of ecosystems.
- Systems ecology is an interdisciplinary field focusing on the study, development, and organization of ecological systems from a holistic perspective.
- Landscape ecology examines processes and relationship in a spatially explicit manner, often across multiple ecosystems or very large geographic areas.
- Evolutionary ecology studies ecology in a way that explicitly considers the evolutionary histories of species and their interactions.
- Political ecology connects politics and economy to problems of environmental control and ecological change.

Ecology can also be sub divided according to the species of interest into fields such as animal ecology, plant ecology, insect ecology, and so on. Another frequent method of subdivision is by biome studied, e.g., Arctic ecology (or polar ecology), tropical ecology, desert ecology, marine ecology, etc. The primary technique used for investigation is often used to subdivide the discipline into groups such as chemical ecology, molecular ecology, field ecology, quantitative ecology, theoretical ecology, and so forth.

Subdivisions of ecology are not mutually exclusive; indeed, very few exist in isolation. Many of them overlap, complement and inform each other. For example, the population ecology of an organism is a consequence of its behavioral ecology and intimately tied to its community ecology. Methods from molecular ecology might inform the study of the population, and all kinds of data are modeled and analyzed using quantitative ecology techniques, often motivated by basic results in theoretical ecology.

FUNDAMENTAL PRINCIPLES

Levels of Organization

Ecology can be studied at a wide range of levels, from large to small scale. These levels of ecological organization, as well as an example of a question ecologists would ask at each level, include:

- Biosphere: " What role does concentration of atmospheric carbon dioxide play in the growth of forests?"
- Region: "How has geological history influenced regional diversity within certain groups of organisms?"
- Landscape: "How do vegetated corridors affect the rate of movement by mammals among isolated fragments?"
- Ecosystem: "How does fire affect nutrient availability in grassland ecosystems?"
- Community: "How does disturbance influence the number of mammal species in African grasslands?"
- Interactions: "What evolutionary benefit do zebras gain by allowing birds to remove parasites?"
- Population: "What factors control zebra populations?"
- Individual Organism: "How do zebras regulate internal water balance?"
- These levels range from broadest to most specific.

Biosphere

For modern ecologists, ecology can be studied at several levels: population level (individuals of the same species in the same or similar environment), biocoenosis level (or community of species), ecosystem level, and biosphere level.

The outer layer of the planet Earth can be divided into several compartments: the hydrosphere (or sphere of water), the lithosphere (or sphere of soils and rocks), and the atmosphere (or sphere of the air). The biosphere (or sphere of life), sometimes described as "the fourth envelope", is all living matter on the planet or that portion of the planet occupied by life. It reaches well into the other three spheres, although there are no permanent inhabitants of the atmosphere. Relative to the volume of the Earth, the biosphere is only the very thin surface layer that extends from 11,000 metres below sea level to 15,000 metres above.

It is thought that life first developed in the hydrosphere, at shallow depths, in the photic zone. (Recently, though, a competing theory has emerged, that life originated around hydrothermal vents in the deeper ocean.) Multicellular organisms then appeared and colonized benthic zones. Photosynthetic organisms gradually produced the chemically unstable oxygen-rich atmosphere that characterizes our planet. Terrestrial life developed later, protected from UV rays by the ozone layer. Diversification of terrestrial species is thought to be increased by the continents drifting apart, or alternately, colliding. Biodiversity is expressed at the ecological level (ecosystem), population level (intraspecific diversity), species level (specific diversity), and genetic level.

The biosphere contains great quantities of elements such as carbon, nitrogen, hydrogen, and oxygen. Other elements, such as phosphorus, calcium, and potassium, are also essential to life, yet are present in smaller amounts. At the ecosystem and biosphere levels, there is a continual recycling of all these elements, which alternate between the mineral and organic states.

Although there is a slight input of geothermal energy, the bulk of the functioning of the ecosystem is based on the input of solar energy. Plants and photosynthetic microorganisms convert light into chemical energy by the process of photosynthesis, which creates glucose (a simple sugar) and releases free oxygen. Glucose thus becomes the secondary energy source that drives the ecosystem. Some of this glucose is used directly by other organisms for energy. Other sugar molecules can be converted to molecules such as amino acids. Plants use some of this sugar, concentrated in nectar, to entice pollinators to aid them in reproduction.

Cellular respiration is the process by which organisms (like mammals) break the glucose back down into its constituents, water and carbon dioxide, thus regaining the stored energy the sun originally gave to the plants. The proportion of photosynthetic activity of plants and other photosynthesizers to the respiration of other organisms determines the specific composition of the Earth's atmosphere, particularly its oxygen level. Global air currents mix the atmosphere and maintain nearly the same balance of elements in areas of intense biological activity and areas of slight biological activity.

Water is also exchanged between the hydrosphere, lithosphere, atmosphere, and biosphere in regular cycles. The oceans are large tanks that store water, ensure thermal and climatic stability, and facilitate the transport of chemical elements thanks to large oceanic currents.

For a better understanding of how the biosphere works, and various dysfunctions related to human activity, American scientists attempted to simulate the biosphere in a small-scale model, called Biosphere 2.

Ecosystem

A central principle of ecology is that each living organism has an ongoing and continual relationship with every other element that makes up its environment. The sum total of interacting living organisms (the biocoenosis) and their non-living environment (the biotope) in an area is termed an *ecosystem*. Studies of ecosystems usually focus on the movement of energy and matter through the system.

Almost all ecosystems run on energy captured from the sun by primary producers via photosynthesis. This energy then flows through the food chains to primary consumers (herbivores who eat and digest the plants), and on to secondary and tertiary consumers (either carnivores or omnivores). Energy is lost to living organisms when it is used by the organisms to do work, or is lost as waste heat.

Matter is incorporated into living organisms by the primary producers. Photosynthetic plants fix carbon from carbon dioxide and nitrogen from atmospheric nitrogen or nitrates present in the soil to produce amino acids. Much of the carbon and nitrogen contained in ecosystems is created by such plants, and is then consumed by secondary and tertiary consumers and incorporated into themselves. Nutrients are usually returned to the ecosystem via decomposition. The entire movement of chemicals in an ecosystem is termed a biogeochemical cycle, and includes the carbon and nitrogen cycle.

Ecosystems of any size can be studied; for example, a rock and the plant life growing on it might be considered an ecosystem. This rock might be within a plain, with many such rocks, small grass, and grazing animals—also an ecosystem. This plain might be in the tundra, which is also an ecosystem (although once they are of this size, they are generally termed ecozones or biomes). In fact, the entire terrestrial surface of the earth, all the matter which composes it, the air that is directly above it, and all the living organisms living within it can be considered as one, large ecosystem.

Ecosystems can be roughly divided into terrestrial ecosystems (including forest ecosystems, steppes, savannas, and so on), freshwater ecosystems (lakes, ponds and rivers), and marine ecosystems, depending on the dominant biotope.

Dynamics and Stability

Ecological factors that affect dynamic change in a population or species in a given ecology or environment are usually divided into two groups: abiotic and biotic.

Abiotic factors are geological, geographical, hydrological, and climatological parameters. A biotope is an environmentally uniform region characterized by a particular set of abiotic ecological factors. Specific abiotic factors include:

- Water, which is at the same time an essential element to life and a milieu
- Air, which provides oxygen, nitrogen, and carbon dioxide to living species and allows the dissemination of pollen and spores
- Soil, at the same time a source of nutriment and physical support
    - Soil pH, salinity, nitrogen and phosphorus content, ability to retain water, and density are all influential
- Temperature, which should not exceed certain extremes, even if tolerance to heat is significant for some species
- Light, which provides energy to the ecosystem through photosynthesis
- Natural disasters can also be considered abiotic

Biocenose, or community, is a group of populations of plants, animals, microorganisms. Each population is the result of procreations between individuals of the same species and cohabitation in a given place and for a given time. When a population consists of an insufficient number of individuals, that population is threatened with extinction; the extinction of a species can approach when all biocenoses composed of individuals of the species are in decline. In small populations, consanguinity (inbreeding) can result in reduced genetic diversity, which can further weaken the biocenose.

Biotic ecological factors also influence biocenose viability; these factors are considered as either intraspecific or interspecific relations.

Intraspecific relations are those that are established between individuals of the same species, forming a population. They are relations of cooperation or competition, with division of the territory, and sometimes organization in hierarchical societies.

Interspecific relations—Interactions between different species—are numerous, and usually described according to their beneficial, detrimental, or neutral effect (for example, mutualism (relation ++) or competition (relation —). The most significant relation is the relation of predation (to eat or to be eaten), which leads to the essential concepts in ecology of food chains (for example, the grass is consumed by the herbivore, itself consumed by a carnivore, itself consumed by a carnivore of larger size). A high predator to prey ratio can have a negative influence on both the predator and prey biocenoses in that low availability of food and high death rate prior to sexual maturity can decrease (or prevent the increase of) populations of each, respectively. Selective hunting of species by humans that leads to population decline is one example of a high predator to prey ratio in action. Other interspecific relations include parasitism, infectious disease, and competition for limited resources, which can occur when two species share the same ecological niche.

The existing interactions between the various living beings go along with a permanent mixing of mineral and organic substances, absorbed by organisms for their growth, their maintenance, and their reproduction, to be finally rejected as waste. These permanent recycling of the elements (in particular carbon, oxygen, and nitrogen) as well as the water are called biogeochemical cycles. They guarantee a durable stability of the biosphere (at least when unchecked human influence and extreme weather or geological phenomena are left aside). This self-regulation, supported by negative feedback controls, ensures the perenniality of the ecosystems. It is shown by the very stable concentrations of most elements of each compartment. This is referred to as homeostasis. The ecosystem also tends to evolve to a state of ideal balance, called the climax, which is reached after a succession of events (for example a pond can become a peat bog).

Spatial Relationships and Subdivisions of Land

Ecosystems are not isolated from each other, but are interrelated. For example, water may circulate between ecosystems by means of a river or ocean current. Water itself, as a liquid medium, even defines ecosystems. Some species, such as salmon or freshwater eels, move between marine systems and fresh-water systems. These relationships between the ecosystems lead to the concept of a *biome.*

A biome is a homogeneous ecological formation that exists over a large region, such as tundra or steppes. The biosphere comprises all of the Earth's biomes—the entirety of places where life is possible—from the highest mountains to the depths of the oceans.

Biomes correspond rather well to subdivisions distributed along the latitudes, from the equator towards the poles, with differences based on the physical environment (for example, oceans or mountain ranges) and the climate. Their variation is generally related to the distribution of species according to their ability to tolerate temperature, dryness, or both. For example, one may find photosynthetic algae only in the *photic* part of the ocean (where light penetrates), whereas conifers are mostly found in mountains.

Though this is a simplification of a more complicated scheme, latitude and altitude approximate a good representation of the distribution of biodiversity within the biosphere. Very generally, the richness of biodiversity (as well for animal as for plant species) is decreasing most rapidly near the equator and less rapidly as one approach the poles.

The biosphere may also be divided into ecozones, which are very well defined today and primarily follow the continental borders. The ecozones are themselves divided into ecoregions, though there is not agreement on their limits.

Ecosystem Productivity

In an ecosystem, the connections between species are generally related to their role in the food chain. There are three categories of organisms:

- *Producers* or *Autotrophs*—Usually plants or cyanobacteria that are capable of photosynthesis but could be other organisms such as the bacteria near ocean vents that are capable of chemosynthesis.
- *Consumers* or *Heterotrophs*—Animals, which can be primary consumers (herbivorous), or secondary or tertiary consumers (carnivorous and omnivores).
- *Decomposers* or *Detritivores*—Bacteria, fungi, and insects which degrade organic matter of all types and restore nutrients to the environment. The producers will then consume the nutrients, completing the cycle.

These relations form sequences, in which each individual consumes the preceding one and is consumed by the one following, in what are called food chains or food networks. In a food network, there will be fewer organisms at each level as one follows the links of the network up the chain, forming a pyramid.

These concepts lead to the idea of biomass (the total living matter in an ecosystem), primary productivity (the increase in organic compounds), and secondary productivity (the living matter produced by consumers and the decomposers in a given time).

These last two ideas are key, since they make it possible to evaluate the carrying capacity—the number of organisms that can be supported by a given ecosystem. In any food network, the energy contained in the level of the producers is not completely transferred to the consumers. The higher up the chain, the more energy and resources are lost. Thus, from a purely energy and nutrient point of view, it is more efficient for humans to be primary consumers (to subsist from vegetables, grains, legumes, fruit, etc.) than to be secondary consumers (consuming herbivores, omnivores, or their products) and still more so than as a tertiary consumer (consuming carnivores, omnivores, or their products). An ecosystem is unstable when the carrying capacity is overrun.

The total productivity of ecosystems is sometimes estimated by comparing three types of land-based ecosystems and the total of aquatic ecosystems. Slightly over half of primary production is estimated to occur on land, and the rest in the ocean.

- The forests (1/3 of the Earth's land area) contain dense biomasses and are very productive.
- Savannas, meadows, and marshes (1/3 of the Earth's land area) contain less dense biomasses, but are productive. These ecosystems represent the major part of what humans depend on for food.
- Extreme ecosystems in the areas with more extreme climates—deserts and semi-deserts, tundra, alpine meadows, and steppes—(1/3 of the Earth's land area) have very sparse biomasses and low productivity
- Finally, the marine and fresh water ecosystems (3/4 of Earth's surface) contain very sparse biomasses (apart from the coastal zones).

Ecosystems differ in biomass (grams carbon per square meter) and productivity (grams carbon per square meter per day), and direct comparisons of biomass and productivity may not be valid. An ecosystem such as that found in taiga may be high in biomass, but slow growing and thus low in productivity. Ecosystems are often compared on the basis of their turnover (production ratio) or turnover time which is the reciprocal of turnover.

Humanity's actions over the last few centuries have seriously reduced the amount of the Earth covered by forests (deforestation), and have increased agro-ecosystems. In recent decades, an increase in the areas occupied by extreme ecosystems has occurred, such as desertification.

Ecological Crisis

Generally, an ecological crisis occurs with the loss of adaptive capacity when the resilience of an environment or of a species or a population evolves in a way unfavourable to coping with perturbations that interfere with that ecosystem, landscape or species survival (Note: The concept of resilience is not universally accepted in ecology, and moreso represents a contingent within the field that take a holist view of the environment. There are also many ecologists that take a reductionistic perspective and that believe that the environment, at base, is indeterministic). It may be that the environment quality degrades compared to the species needs, after a change in an abiotic ecological factor (for example, an increase of temperature, less significant rainfalls). It may be that the environment becomes unfavourable for the survival of a species (or a population) due to an increased pressure of predation (for example overfishing). Lastly, it may be that the situation becomes unfavourable to the quality of life of the species (or the population) due to a rise in the number of individuals (overpopulation).

Ecological crises vary in length and severity, occurring within a few months or taking as long as a few million years. They can also be of natural or anthropic origin. They may relate to one unique species or to many species, as in an Extinction event. Lastly, an ecological crisis may be local (as an oil spill) or global (a rise in the sea level due to global warming).

According to its degree of endemism, a local crisis will have more or less significant consequences, from the death of many individuals to the total extinction of a species. Whatever its origin, disappearance of one or several species often will involve a rupture in the food chain, further impacting the survival of other species.

In the case of a global crisis, the consequences can be much more significant; some extinction events showed the disappearance of more than 90% of existing species at that time. However, it should be noted that the disappearance of certain species, such as the dinosaurs, by freeing an ecological niche, allowed the development and the diversification of the mammals. An ecological crisis thus paradoxically favoured biodiversity.

Sometimes, an ecological crisis can be a specific and reversible phenomenon at the ecosystem scale. But more generally, the crises impact will last. Indeed, it rather is a connected series of events, that occur till a final point. From this stage, no return to the previous stable state is possible, and a new stable state will be set up gradually.

Lastly, if an ecological crisis can cause extinction, it can also more simply reduce the quality of life of the remaining individuals. Thus, even if the diversity of the human population is sometimes considered threatened, few people envision human disappearance at short span. However, epidemic diseases, famines, impact on health of reduction of air quality, food crises, reduction of living space, accumulation of toxic or non degradable wastes, threats on keystone species (great apes, panda, whales) are also factors influencing the well-being of people.

Due to the increases in technology and a rapidly increasing population, humans have more influence on their own environment than any other ecosystem engineer.

SCIENCE OF EARTH

Earth science (also known as geoscience, the geosciences or the Earth sciences), is an all-embracing term for the sciences related to the planet Earth. It is arguably a special case in planetary science, the Earth being the only known life-bearing planet. There are both reductionist and holistic approaches to Earth science.

There are four major disciplines in earth sciences, namely geography, geology, geophysics and geodesy. These major disciplines use physics, chemistry, biology, chronology and mathematics to build a quantitative understanding of the principal areas or *spheres* of the Earth system.

Fields of Study

The following fields of science are generally categorized within the geosciences:

- Geology describes the rocky parts of the Earth's crust (or lithosphere) and its historic development. Major subdisciplines are mineralogy and petrology, geochemistry, geomorphology, paleontology, stratigraphy, structural geology, engineering geology and sedimentology.
- Geophysics and Geodesy investigate the figure of the Earth, its reaction to forces and its magnetic and gravity fields. Geophysicists explore the Earth's core

and mantle as well as the tectonic and seismic activity of the lithosphere.

- Soil science covers the outermost layer of the Earth's crust that is subject to soil formation processes (or pedosphere). Major subdisciplines include edaphology and pedology.
- Oceanography and hydrology (includes limnology) describe the marine and freshwater domains of the watery parts of the Earth (or hydrosphere). Major subdisciplines include hydrogeology and physical, chemical, and biological oceanography.
- Glaciology covers the icy parts of the Earth (or cryosphere).
- Atmospheric sciences cover the gaseous parts of the Earth (or atmosphere) between the surface and the exosphere (about 1000 km). Major subdisciplines are meteorology, climatology, atmospheric chemistry and atmospheric physics.
- A very important linking sphere is the biosphere, the study of which is biology. The biosphere consists of all forms of life, from single-celled organisms to pine trees to people. The interactions of Earth's other spheres - lithosphere/geosphere, hydrosphere, atmosphere and/ or cryosphere and pedosphere—create the conditions that can support life.

Earth's Interior

Plate tectonics, mountain ranges, volcanoes, and earthquakes are geological phenomena that can be explained in terms of energy transformations in the Earth's crust.

Beneath the Earth's crust lies the mantle which is heated by the radioactive decay of heavy elements. The mantle is not quite solid and consists of magma which is in a state of semi-perpetual convection. This convection process causes the lithospheric plates to move, albeit slowly. The resulting process is known as plate tectonics.

Plate tectonics might be thought of as the process by which the earth resurfaces itself. Through a process called *spreading ridges* (or seafloor spreading), the earth creates new crust by allowing magma underneath the lithosphere to come to the surface where it cools and solidifies—becoming new crust, and through a process called subduction, excess crust is pushed underground—beneath the rest of the lithosphere—where it comes into contact with magma and melts—rejoining the mantle from which it originally came.

Areas of the crust where new crust is created are called *divergent boundaries,* and areas of the crust where it is brought back into the earth are called *convergent boundaries*. Earthquakes result from the movement of the lithospheric plates, and they often occur near covergent boundaries where parts of the crust are forced into the earth as part of subduction.

Volcanoes result primarily from the melting of subducted crust material. Crust material that is forced into the Asthenosphere melts, and some portion of the melted material becomes light enough to rise to the surface—giving birth to volcanoes.

Earth's Electromagnetic Field

An electromagnet is a magnet that is created by a current that flows around a soft-iron core. Earth has a soft iron core surrounded by semi-liquid materials from the mantle that move in continuous currents around the core; therefore, the earth is an electromagnet. This is referred to as the dynamo theory of Earth's magnetism. The fact that Earth is an electromagnet helps it maintain an atmosphere suitable for life.

Atmosphere

Earth is blanketed by an atmosphere consisting of 78.0% nitrogen, 20.9% oxygen, and 1% Argon. The atmosphere has five layers: troposphere, stratosphere, mesosphere, thermosphere, and exosphere; and 75% of the atmosphere's gases are in the bottom-most layer, the troposphere.

The magnetic field created by mantle's internal motions produces the magnetosphere which protects the Earth's atmosphere from the solar wind. It is theorized that the solar wind would strip away earth's atmosphere in a few million years were it not for the Earth's electromagnet. And since earth is 4.5 billion years old, earth would not have an atmosphere by now if there were no magnetosphere.

The atmosphere is composed of 78% nitrogen and 21% oxygen. The remaining one per cent contains small amounts of other gases including CO_2 and water vapours. Water vapours and CO_2 allow the Earth's atmosphere to catch and hold the sun's energy through a phenomenon called the greenhouse effect. This allows earth's surface to be warm enough to have liquid water and support life.

In addition to storing heat, the atmosphere also protects living organisms by shielding the Earth's surface from cosmic rays. Note that the level of protection is high enough to prevent cosmic rays from destroying all life on Earth, yet low enough to aid the mutations that have an important role in pushing forward diversity in the biosphere.

Methodology

Like all other scientists, Earth scientists apply the scientific method. They formulate hypotheses after observing events and gathering data about natural phenomena, and then they test hypotheses from such data.

A contemporary idea within earth science is uniformitarianism. Uniformitarianism says that "ancient geologic features are interpreted by understanding active processes that are readily observed". Simply stated, this means that features of the Earth can be explained by the actions of gradual processes operating over long periods of time; for example, a mountain need not be thought of as having been created in a moment, but instead it may be seen as the result of continuous subduction, causing magma to rise and form continental volcanic arcs.

Earth's Spheres

Earth science generally recognizes four spheres, the lithosphere, the hydrosphere, the atmosphere, and the biosphere; these correspond to rocks, water, air, and life. Some practitioners include, as part of the spheres of the Earth, the cryosphere (corresponding to ice) as a distinct portion of the hydrosphere, as well as the pedosphere (corresponding to soil) as an active and intermixed sphere.

Environment (Biophysical)

The biophysical environment is the symbiosis between the physical environment and the biological life forms within the environment, and includes all variables that comprise the Earth's biosphere. The biophysical environment can be divided into two categories: the natural environment and the built environment, with some overlap between the two. Following the industrial revolution, the built environment has become an increasingly significant part of the Earth's environment.

Constituents

The scope of the biophysical environment is all that contained in the biosphere, which is that part of the Earth in which all life occurs. Ecosystems, of which there are numerous types and are a defined part of the biosphere, collectively make up the whole of the biosphere. Within an ecosystem there are habitats in which an organism (including humans) exists. At its most natural, an environment would lack any effects of human activity, although the scale of this activity is such that all areas of the Earth have had at least some influence by humans. At the other end of the scale is the built environment and in some cases it has the biotic component that is virtually absent.

The biophysical environment can vary in scale from microscopic to global in extent. They can also be subdivided according to their attributes. Some examples may be the Marine environment, the atmospheric environment and the terrestrial environment.

Environmental Science

Environmental science is the study of the interactions within the biophysical environment. Part of this scientific discipline is the investigation of the effect of human activity on the environment. Ecology, a sub-discipline of biology and a part of environmental sciences, is often mistaken as a study of human induced effects on the environment. Environmental studies is a broader academic discipline that is the systematic study of interaction of humans with their environment. It is a broad field of study that includes the natural environment, built environments and social environments.

Environmentalism is a broad social and philosophical movement that, in a large part, seeks to minimise or eliminate the effect of human activity on the biophysical environment. The issues of concern for environmentalists usually relate to the natural environment with the more important ones being climate change, species extinction, pollution and old growth forest loss.

5
Atmospheric Environment

The outgassings of the Earth was stripped away by solar winds early in the history of the planet till a steady state was established, the first atmosphere. Based on today's volcanic evidence, this atmosphere would have contained 80% water vapor, 10% carbon dioxide, 5 to 7% hydrogen-sulfide, and smaller amounts of nitrogen, carbon monoxide, hydrogen, methane and inert gases.

A major rainfall lead to the buildup of a vast ocean, enriching the other agents, first carbon dioxide and later nitrogen and inert gases. A major part of carbon dioxide exhalations were soon dissolved in water and built up carbonate sediments.

Second Atmosphere

As early as 3.8 Billion years ago, water related sediments have been found. About 3.4 billion years ago, nitrogen was the major part of the then stable second atmosphere. An influence of life has to be taken into account rather soon since hints on early life forms are to be found as early as 3.5 billion years ago. The fact that this is not in line with the — compared to today 30% lower — solar radiance of the early sun has been described as the faint young Sun paradox.

The geological record, however, shows a continually relatively warm surface during the complete early temperature record of the earth with the exception of one cold glacial phase about 2.4 billion years ago. Sometime during the late

Archean Era an oxygen containing atmosphere began to develop from photosynthesizing algae. The early basic carbon isotopy is very much in line with what is found today and the fundamental features of the carbon cycle were established as early as 4 billion years ago. As Jan Veizer assumed that not only did we have life as far back as we had rocks, but there was as much life then as today and the fundamental features of the carbon cycle were established as early as 4 billion years ago.

Third Atmosphere

The accretion of continents about 3.5 billion years ago added plate tectonics, constantly rearranging the continents and also shaping long-term climate evolution by allowing the transfer of carbon dioxide to large land-based carbonate storages. Free oxygen did not exist until about 1.7 Ga ago and this can be seen with the development of the red beds and the end of the banded iron formations. This signifies a shift from a reducing atmosphere to an oxidising atmosphere. O_2 showed major ups and downs until reaching a steady state of more than 15%. The following time span was the Phanerozoic, during which oxygen-breathing metazoan life forms began to appear.

Climate during Geological Ages

Precambrian Climate

In the first three quarters of the Earth's history, only one major glaciation is to be found in the geological record. Since about 950 million years ago, the Earth's climate has varied regularly between large-scale or just polar cap wide glaciation and extensively tropical climates. The time scale for this variation is roughly 140 million years and may be related to Earth's motion into and out of galactic spiral arms and compared to the previous time, significantly reduced solar wind.

The climate of the late Precambrian showed some major glaciation events spreading over much of the earth. At this time the continents were bunched up in the Rodinia supercontinent. Massive deposits of tillites are found and anomalous isotopic signatures are found, which gave rise to the Snowball Earth hypothesis. As the Proterozoic Eon drew to a close, the Earth started to warm up. By the dawn of the Cambrian and the Phanerozoic, life forms were abundant in the Cambrian explosion with average global temperatures of about 22°C.

Phanerozoic Climate

Major drivers for the preindustrial ages have been variations of the sun, volcanic ashes and exhalations, relative movements of the earth towards the sun and tectonically induced effects as for major sea currents, watersheds and ocean oscillations. Especially the early Phanerozoic shows a decoupling of a rather high carbon dioxide content and significant global glaciation. Royer et al. 2004 found a climate sensitivity for the rest of the Phanerozoic which was calculated to be similar to today's modern range of values.

The difference in global mean temperatures between a fully glacial Earth and an ice free Earth is estimated at approximately 10°C, though far larger changes would be observed at high latitudes and smaller ones at low latitudes. One requirement for the development of large scale ice sheets seems to be the arrangement of continental land masses at or near the poles. The constant rearrangement of continents by plate tectonics can also shape long-term climate evolution. However, the presence or absence of land masses at the poles is not sufficient to guarantee glaciations or exclude polar ice caps. Evidence exists of past warm periods in Earth's climate when polar land masses similar to Antarctica were home to deciduous forests rather than ice sheets.

The relatively warm local minimum between Jurassic and Cretaceous goes along with widespread tectonic activity, e.g. the breakup of supercontinents.

Superimposed on the long-term evolution between hot and cold climates have been many short-term fluctuations in climate similar to, and sometimes more severe than, the varying glacial and interglacial states of the present ice age. Some of the most severe fluctuations, such as the Paleocene-Eocene Thermal Maximum, may be related to rapid climate changes due to sudden collapses of natural methane clathrate reservoirs in the oceans.

A similar, single event of induced severe climate change after a meteorite impact has been proposed as reason for the Cretaceous-Tertiary extinction event. Other major thresholds are the Permian-Triassic, and Ordovician-Silurian extinction events with various reasons suggested.

Quaternary Sub-Era

The Quaternary sub-era includes the current climate. There has been a cycle of ice ages for the past 2.2–2.1 million years (starting before the Quaternary in the late Neogene Period).

Note in the graphic on the right the strong 120,000-year periodicity of the cycles, and the striking asymmetry of the curves. This asymmetry is believed to result from complex interactions of feedback mechanisms. It has been observed that ice ages deepen by progressive steps, but the recovery to interglacial conditions occurs in one big step.

Controlling Factors

Short-Term (10 to 10 years)

Geologically short-term (<120,000 year) temperatures are believed to be driven by orbital factors (see Milankovitch cycles) amplified by changes in greenhouse gases. The arrangements of land masses on the Earth's surface are believed to influence the effectiveness of these orbital forcing effects.

Medium-Term (10 to 10 years)

Continental drift affects the thermohaline circulation, which transfers heat between the equatorial regions and the poles, as does the extent of polar ice coverage.

The timing of ice ages throughout geologic history is in part controlled by the position of the continental plates on the surface of the Earth. When landmasses are concentrated near the polar regions, there is an increased chance for snow and ice to accumulate. Small changes in solar energy can tip the balance between summers in which the winter snow mass completely melts and summers in which the winter snow persists until the following winter. See the web site Paleomap Project for images of the polar landmass distributions through time.

Comparisons of plate tectonic continent reconstructions and paleoclimatic studies show that the Milankovitch cycles have the greatest effect during geologic eras when landmasses have been concentrated in polar regions, as is the case today. Today, Greenland, Antarctica, and the northern portions of Europe, Asia, and North America are situated such that a minor change in solar energy will tip the balance between year-round snow/ice preservation and complete summer melting. The presence of snow and ice is a well-understood positive feedback mechanism for climate. The Earth today is considered to be prone to ice age glaciations.

Another proposed factor in long term temperature change is the Uplift-Weathering Hypothesis, first put forward by T. C. Chamberlin in 1899 and later independently proposed in 1988 by Maureen Raymo and colleagues, where upthrusting mountain ranges expose minerals to weathering resulting in their chemical conversion to carbonates thereby removing CO_2 from the atmosphere and cooling the earth. Others have proposed similar effects due to changes in average water table levels and consequent changes in sub-surface biological activity and PH levels.

Long-Term (10 to 10 years)

It has been proposed that long term galactic motions of the sun have a major influence on earth's climate. There are two principal motions, the first and most significant is the orbit of the sun around the galactic centre with a period of the order of 240 million years. Since this period is different from the rotation period of the galactic spiral arms, the sun, and the earth with it, will periodically pass through the arms (estimates of the period are uncertain and vary from 143 million years to 176 million years). The second is an oscillatory bobbing motion, similar to a floating buoy, which will periodically take the sun through the galactic disc. The period of this bobbing motion is 67 million years, so a pass through the galactic plane will occur every 33 million years. The causal link between these galactic motions and climate is postulated to be the effect that entering a denser region of the galaxy will have on increasing the Cosmic Ray Flux (CRF). This theory has been criticised, both for overstating the correlation with CRF and for failing to propose a believable mechanism that would allow CRF to drive temperature. The claims by Henrik Svensmark that CRF also strongly affects short term climate changes is even more controversial and has been challenged by many.

It has also been suggested that there is some correlation between these galactic cycles and geological periods. The reason for this is postulated to be that the earth experiences many more impact events while passing through high density regions of the galaxy. Both the climate changes and sudden impacts may cause, or contribute to, extinction events.

Very Long-Term (10 years or more)

Jan Veizer and Nir Shaviv have proposed the interaction of cosmic rays, solar wind and the various magnetic fields to explain the long term evolution of earths climate. According to Shaviv, the early sun had emitted a stronger solar wind with a protective effect against cosmic rays. In that early age, a moderate greenhouse effect comparable to today's would have been sufficient to explain an ice free earth and the faint young sun paradox. The solar minimum around 2.4 billion years ago is consistent with an established Cosmic Ray Flux modulation by a variable star formation rate in the Milky Way and there is also a hint of an extinction event at this time. Within the last billion years the solar wind has significantly diminished. It is only within this more recent time that passages of the heliosphere through the spiral arms of the galaxy have been able to gain a strong and regularly modulating influence as described above.

Over the very long-term the energy output of the sun has gradually increased, on the order of 5% per billion (10^9) years, and will continue to do so until it reaches the end of its current phase of stellar evolution.

Atmospheric Physics

Atmospheric Physics is the application of physics to the study of the atmosphere. Atmospheric physicists attempt to model Earth's atmosphere and the atmospheres of the other planets using fluid flow equations, chemical models, radiation balancing, and energy transfer processes in the atmosphere (as well as how these tie in to other systems such as the oceans). In order to model weather systems, atmospheric physicists employ elements of scattering theory, wave propagation models, cloud physics, statistical mechanics and spatial statistics which are highly mathematical and related to physics. It has close links to meteorology and climatology and also covers the design and construction of instruments for studying the atmosphere and the interpretation of the data they provide, including remote sensing instruments. At the dawn of the space age and the introduction of sounding rockets, aeronomy became a subdiscipline concerning the upper layers of the atmosphere, where dissociation and ionization are important.

Remote Sensing

Remote sensing is the small or large-scale acquisition of information of an object or phenomenon, by the use of either recording or real-time sensing device(s) that is not in physical or intimate contact with the object (such as by way of aircraft, spacecraft, satellite, buoy, or ship). In practice, remote sensing is the stand-off collection through the use of a variety of devices for gathering information on a given object or area which gives more information than sensors at individual sites might convey. Thus, Earth observation or weather satellite collection platforms, ocean and atmospheric observing weather buoy platforms, monitoring of a pregnancy via ultrasound, Magnetic Resonance Imaging (MRI), Positron Emission Tomography (PET), and space probes are all examples of remote sensing. In modern usage, the term generally refers to the use of imaging sensor technologies including but not limited to the use of instruments aboard aircraft and spacecraft, and is distinct from other imaging-related fields such as medical imaging.

There are two kinds of remote sensing. Passive sensors detect natural radiation that is emitted or reflected by the object or surrounding area being observed. Reflected sunlight is the most common source of radiation measured by passive sensors. Examples of passive remote sensors include film photography, infra-red, charge-coupled devices, and radiometers. Active collection, on the other hand, emits energy in order to scan objects and areas whereupon a passive sensor then detects and measures the radiation that is reflected or backscattered from the target. RADAR is an example of active remote sensing where the time delay between emission and return is measured, establishing the location, height, speed and direction of an object.

Remote sensing makes it possible to collect data on dangerous or inaccessible areas. Remote sensing applications include monitoring deforestation in areas such as the Amazon Basin, the effects of climate change on glaciers and Arctic and Antarctic regions, and depth sounding of coastal and ocean depths. Military collection during the cold war made use of stand-off collection of data about dangerous border areas. Remote sensing also replaces costly and slow data collection on the ground, ensuring in the process that areas or objects are not disturbed.

Orbital platforms collect and transmit data from different parts of the electromagnetic spectrum, which in conjunction with larger scale aerial or ground-based sensing and analysis, provides researchers with enough information to monitor trends such as El Niño and other natural long and short term phenomena. Other uses include different areas of the earth sciences such as natural resource management, agricultural fields such as land usage and conservation, and national security and overhead, ground-based and stand-off collection on border areas.

Radiation

The sun emits radiation as a variety of wavelengths. Visible light has wavelengths between 0.4 and 0.7 micrometers. Shorter wavelengths are known as the ultraviolet (UV) part of the spectrum, while longer wavelengths are grouped into the infrared portion of the spectrum. Ozone is most effective in absorbing radiation around 0.25 micrometers, where UV-c rays lie in the spectrum. This increases the temperature of the nearby stratosphere. Snow reflects 88% of UV rays, while sand reflects 12%, and water reflects only 4% of incoming UV radiation. The more glancing the angle is between the atmosphere and the sun's rays, the more likely that energy will be reflected or absorbed by the atmosphere.

Cloud Physics

Cloud physics is the study of the physical processes that lead to the formation, growth and precipitation of clouds. Clouds are composed of microscopic droplets of water (warm clouds), tiny crystals of ice, or both (mixed phase clouds). Under suitable conditions, the droplets combine to form precipitation, where they may fall to the earth. The precise mechanics of how a cloud forms and grows is not completely understood, but scientists have developed theories explaining the structure of clouds by studying the microphysics of individual droplets. Advances in radar and satellite technology have also allowed the precise study of clouds on a large scale.

Atmospheric Electricity

Atmospheric electricity is the regular diurnal variations of the Earth's atmospheric electromagnetic network (or, more broadly, any planet's electrical system in its layer of gases). The Earth's surface, the ionosphere, and the atmosphere is known as the global atmospheric electrical circuit. Lightning discharges 30,000 amperes, at up to 100 million volts, and emits light, radio waves, x-rays and even gamma rays. Plasma temperatures in lightning can approach 28,000 kelvins and electron densities may exceed $10^{24}/m^3$.

Atmospheric Tide

The largest-amplitude atmospheric tides are mostly generated in the troposphere and stratosphere when the atmosphere is periodically heated as water vapour and ozone absorb solar radiation during the day. The tides generated are then able to propagate away from these source regions and ascend into the mesosphere and thermosphere. Atmospheric tides can be measured as regular fluctuations in wind, temperature, density and pressure. Although atmospheric tides share much in common with ocean tides they have two key distinguishing features:

(i) Atmospheric tides are primarily excited by the Sun's heating of the atmosphere whereas ocean tides are primarily excited by the Moon's gravitational field. This means that most atmospheric tides have periods of oscillation related to the 24-hour length of the solar day whereas ocean tides have longer periods of oscillation related to the lunar day (time between successive lunar transits) of about 24 hours 51 minutes.

(ii) Atmospheric tides propagate in an atmosphere where density varies significantly with height. A consequence of this is that their amplitudes naturally increase exponentially as the tide ascends into progressively more rarefied regions of the atmosphere (for an explanation of this phenomenon, see below). In contrast, the density of the oceans varies only slightly with depth and so there the tides do not necessarily vary in amplitude with depth.

Note that although solar heating is responsible for the largest-amplitude atmospheric tides, the gravitational fields of the Sun and Moon also raise tides in the atmosphere, with the lunar gravitational atmospheric tidal effect being significantly greater than its solar counterpart.

At ground level, atmospheric tides can be detected as regular but small oscillations in surface pressure with periods of 24 and 12 hours. Daily pressure maxima occur at 10 a.m. and 10 p.m. local time, while minima occur at 4 a.m. and 4 p.m. local time. The absolute maximum occurs at 10 a.m. while the absolute minimum occurs at 4 p.m. However, at greater heights the amplitudes of the tides can become very large. In the mesosphere (heights of ~ 50 - 100 km) atmospheric tides can reach amplitudes of more than 50 m/s and are often the most significant part of the motion of the atmosphere.

Aeronomy

Aeronomy is the science of the upper region of the atmosphere, where dissociation and ionization are important. The term aeronomy was introduced by Sydney Chapman in 1960. Today, the term also includes the science of the corresponding regions of the atmospheres of other planets. Research in aeronomy requires access to balloons, satellites, and sounding rockets which provide valuable data about this region of the atmosphere. Atmospheric tides play an important role in interacting with both the lower and upper atmosphere. Amongst the phenomena studied are upper-atmospheric lightning discharges, such as luminous events called red sprites, sprite halos, blue jets, and elves.

Centres of Research

In the UK, atmospheric studies are underpinned by the Met Office, the Natural Environment Research Council and the Science and Technology Facilities Council. Divisions of the U.S. National Oceanic and Atmospheric Administration (NOAA) oversee research projects and weather modeling involving atmospheric physics. The US National Astronomy and Ionosphere Center also carries out studies of the high atmosphere. In Belgium, the Belgian Institute for Space Aeronomy studies the atmosphere and outer space.

Faint Young Sun Paradox

The faint young Sun paradox or problem describes the apparent contradiction between observations of liquid water early in the Earth's history and the astrophysical expectation that the Sun's output would be only 70% as intense during that epoch as it is during the modern epoch. The issue was raised by astronomers Carl Sagan and George Mullen in 1972.

According to the Standard Solar Model, stars similar to the Sun should gradually brighten over their life time.

However, with the predicted brightness 4 billion (10^9) years ago and with greenhouse gas concentrations the same as are current for the modern Earth, any liquid water exposed to the surface would freeze. The geological record however shows a continually relatively warm surface during the complete early Temperature record of the earth with the exception of a cold phase about 2.4 billion years ago. As early as 3.8 Billion years ago, water related sediments have been found . Hints on early life forms are as early as 3.5 Billion years, the basic carbon isotopy is very much in line with what is found today . A regular change between ice ages and warm periods is only to be found since one billion years.

Explanations of the paradox take into account either greenhouse effects or astrophysical deliberations, or a combination of the two.

Greenhouse Hypothesis

Earth's atmosphere may have contained more greenhouse gases early on. Carbon dioxide concentrations may have been higher without plant photosynthesis converting it to oxygen. Methane, a very active greenhouse gas which reacts with oxygen to produce carbon dioxide, may have been more prevalent as well.

Following the accretion of continents after about 1 billion years, Walter and others assume a non-biological version of the carbon cycle to provide a negative temperature feedback. Carbon dioxide dissolves in liquid water and combines with metal ions derived from silicate weathering to produce carbonates. During ice ages this part of the cycle would shut down. Volcanic carbon emissions would restart a warming cycle due to the greenhouse effect.

There may have been a number of periods when the Earth's oceans froze over completely (Snowball Earth hypothesis). The most recent period may have been ~630 million years ago. Afterwards the Cambrian explosion of new multicellular life forms started.

Astronomical and Geochemical Evidence and Hypothesis

Nir Shaviv uses climatical influences of Solar Wind and — along the hypothesis of Henrik Svensmark — a cooling effect of Cosmic Rays to explain the paradox.

According to Shaviv, the early sun had emitted a stronger solar wind with a protective effect against cosmic rays. In that early age, a moderate greenhouse effect comparable to today's would have been sufficient to explain an ice free earth. The minimum around 2,4 billion years goes along with a cosmic ray flux modulation by a variable star formation rate in the Milky Way.

The reduced solar impact later results into a stronger impact of cosmic ray flux (CRF). Detailed Measurements of the CRF based on Meteorite evidence are in line with the global ice ages of the current eon. Besides, isotopic evidence from the geochemical record is able to trace either an of the cosmic ray flux(14C and 10Be) and climate change based on ä18 Oxygen measurements of sediments. Either are in line with climatical variations till the most recent past.

Oxygen Catastrophe

The Oxygen Catastrophe was a massive environmental change believed to have happened during the Siderian period at the beginning of the Paleoproterozoic era of the Precambrian, about 2.4 billion years ago. It is also called the Oxygen Crisis, Oxygen Revolution, or The Great Oxidation.

When evolving lifeforms developed oxyphotosynthesis about 3.5 billion years ago, molecular oxygen was initially produced in limited quantities. With time, this oxygen accumulated and eventually caused an ecological crisis to the biodiversity of the time, as oxygen was toxic to the microscopic anaerobic organisms dominant then.

However, this transforming change also provided a new opportunity for biological diversification, as well as tremendous changes in the nature of chemical interactions between rocks, sand, clay, and other geological substrates and the Earth's air, oceans, and other surface waters. Despite natural recycling of organic matter, life had remained energetically limited until the widespread availability of oxygen. This breakthrough in metabolic evolution greatly increased the free energy supply to living organisms, having a truly global environmental impact.

Time Lag

There was a lag of about 300 million years between the time oxygen production from photosynthetic organisms started (about 2.8 billion years ago), and the time of the Oxygen Catastrophe's geologically rapid increase in atmospheric oxygen (about 2.5 - 2.4 billion years ago). There are a number of hypotheses to explain this time lag:

Tectonic Trigger

One phenomenon that explains this lag is that the oxygen increase had to await tectonically driven changes in the Earth's 'anatomy,' including the appearance of shelf seas where reduced organic carbon could reach the sediments and be buried. Also, the newly produced oxygen was first consumed in various chemical reactions in the oceans, primarily with iron. Evidence for this phenomenon is found in older rocks that contain massive banded iron formations that were apparently laid down as this iron and oxygen first combined; most of the planet's commercial iron ore deposits are in these deposits. But these chemical phenomena do not seem to account for the lag completely.

Nickel Famine

Photosynthetic organisms were a source of methane, which was also a big trap for molecular oxygen, because oxygen readily oxidizes methane to carbon dioxide (CO_2) and water in the presence of UV radiation. Modern methanogens require nickel as an enzyme cofactor. As the Earth's crust cooled, the supply of nickel from volcanoes was reduced and less methane was produced allowing oxygen to dominate the atmosphere. From 2.7 to 2.4 billion years ago, the levels of nickel deposited declined steadily; it was originally 400 times today's levels.

Bistability

A 2006 (bistability) theory to explain the 300-million-year lag comes from a mathematical model of the atmosphere which recognizes that UV shielding decreases the rate of methane oxidation once oxygen levels are sufficient to support the formation of an ozone layer. This explanation proposes a system with two steady states, one with lower (0.02%) atmospheric oxygen content, and the other with higher (21% or more) oxygen content. The Great Oxidation can then be understood as a switch between lower and upper stable steady states.

Hydrogen Leakage

Another factor in the delay in atmospheric oxygen enrichment may have been photosynthetic production of molecular hydrogen which, as it formed, got into the atmosphere and was slowly lost to space.

Snowball Earth

Snowball Earth refers to hypotheses regarding paleoclimatic global-scale glaciation, claiming that the Earth's surface was nearly or entirely frozen at some points in its past. The occurrence of Snowball (or Slushball) Earths remains controversial. Proponents claim it best explains sedimentary deposits generally regarded as of glacial origin at tropical latitudes and other enigmatic features of the geological record. Opponents contest the implications of that geological evidence and the geophysical feasibility of an ice- or slush-covered ocean.

History

Sir Douglas Mawson (1882–1958), an Australian geologist and Antarctic explorer, spent much of his career studying the Neoproterozoic stratigraphy of South Australia where he identified thick and extensive glacial sediments and late in his career speculated on the possibility of global glaciation. Mawson's ideas of global glaciation, however, were based on the mistaken assumption that the geographic position of Australia, and that of other continents where low-latitude glacial deposits are found, has remained constant through time. With the advancement of the continental drift hypothesis, and eventually plate tectonic theory, came an easier explanation for the glaciogenic sediments — they were deposited at a point in time when the continents were at higher latitudes. In 1964 the idea of global-scale glaciation reemerged when W. Brian Harland published a paper in which he presented palaeomagnetic data showing that glacial tillites in Svalbard and Greenland were deposited at tropical latitudes. From this palaeomagnetic data, and the sedimentological evidence that the glacial sediments interrupt successions of rocks commonly associated with tropical to temperate latitudes, he argued for an ice age that was so extreme that it resulted in the deposition of marine glacial rocks in the tropics.

In the 1960s, Mikhail Budyko, a Russian climatologist, developed a simple energy-balance climate model to investigate the effect of ice cover on global climate. Using this model, Budyko found that if ice sheets advanced far enough out of the polar regions a feedback ensued where the increased reflectiveness (albedo) of the ice led to further cooling and the formation of more ice until the entire Earth was covered in ice and stabilized in a new ice-covered equilibrium. While Budyko's model showed that this ice-albedo *stability* could happen, he concluded that it had never happened, because his model offered no way to escape from such a scenario.

The term "Snowball Earth" was coined by Joseph Kirschvink, a professor of geobiology at the California Institute of Technology, in a short paper published in 1992 within a lengthy volume concerning the biology of the Proterozoic eon. The major contributions from this work were: (1) the recognition that the presence of banded iron formations is consistent with such a glacial episode and (2) the introduction of a mechanism with which to escape from an ice-covered Earth — the accumulation of CO_2 from volcanic outgassing leading to an ultra-greenhouse effect.

Interest in the Snowball Earth increased dramatically after Paul F. Hoffman, professor of geology at Harvard University, and coauthors applied Kirschvink's ideas to a succession of Neoproterozoic sediments in Namibia, elaborated upon the hypothesis by incorporating such observations as the occurrence of cap carbonates, and published their results in the journal *Science* in 1998.

Currently, aspects of the hypothesis remain controversial and it is being debated under the auspices of the International Geoscience Programme (IGCP) Project 512: Neoproterozoic Ice Ages.

Evidence

The Snowball Earth hypothesis was originally devised to explain the apparent presence of glaciers at tropical latitudes. Modelling suggested that once glaciers spread to within 30° of the equator, an ice-albedo feedback would result in the ice rapidly advancing to the equator.* Therefore, the presence of glacial deposits seemingly within the tropics appeared to point to global ice cover.

Critical to an assessment of the validity of the theory, therefore, is an understanding of the reliability and significance of the evidence that led to the belief that ice ever reached the tropics. This evidence must prove two things:

1. that a bed contains sedimentary structures that could only have been created by glacial activity;
2. that the bed lay within the tropics when it was deposited.

   During a period of global glaciation, it must also be demonstrated that

3. glaciers were active at different global locations at the same time, and that no other deposits of the same age are in existence.

This latter point is very difficult to prove. Before the Ediacaran, the biostratigraphic markers usually used to correlate rocks are absent; therefore there is no way to prove that rocks in different places across the globe were deposited at the same time. The best we can do is to estimate the age of the rocks using radiometric methods, which are rarely accurate to better than ± a million years or so.

The first two points are often the source of contention on a case-to-case basis. Many glacial features can also be created by non-glacial means, and estimating the latitude of landmasses even as little as 200 million years ago can be riddled with difficulties.

Palaeomagnetism

The Snowball Earth hypothesis was first posited in order to explain what were then considered to be glacial deposits near the equator.

Since continents drift with time, ascertaining their position at a given point in history is far from trivial. In addition to considerations of how the continents would have fitted together, the latitude at which a rock was deposited can be constrained by palaeomagnetism.

When sedimentary rocks form, magnetic minerals within them tend to align themselves with the Earth's magnetic field.

Through the precise measurement of this palaeomagnetism, it is possible to estimate the latitude (but not the longitude) where the rock matrix was deposited. Paleomagnetic measurements have indicated that some sediments of glacial origin in the Neoproterozoic rock record were deposited within 10 degrees of the equator, although the accuracy of this reconstruction is in question. This palaeomagnetic location of apparently glacial sediments (such as dropstones) has been taken to suggest that glaciers extended to sea-level in the tropical latitudes. It is not clear whether this can be taken to imply a global glaciation, or the existence of localised, possibly land-locked, glacial regimes. Others have even suggested that most data do not constrain any glacial deposits to within 25° of the equator.

Skeptics suggest that the palaeomagnetic data could be corrupted if the Earth's magnetic field was substantially different from today's. Depending on the rate of cooling of the Earth's core, it is possible that during the Proterozoic, its magnetic field did not approximate a dipolar distribution, with a North and South pole roughly aligning with the planet's axis as they do today. Instead, a hotter core may have circulated more vigorously and given rise to 4, 8 or more poles. Paleomagnetic data would then have to be re-interpreted as particles could align pointing to a 'West Pole' rather than the North Pole.

Another weakness of reliance on palaeomagnetic data is the difficulty in determining whether the magnetic signal recorded is original, or whether it has been reset by later activity. For example, a mountain-building orogeny releases hot water as a by-product of metamorphic reactions; this water can circulate to rocks thousands of km away and reset their magnetic signature. This makes the authenticity of rocks older than a few million years difficult to determine without painstaking mineralogical observations.

There is currently only one deposit, the Elatina deposit of Australia, that was indubitably deposited at low latitudes; its depositional date is well constrained, and the signal is demonstrably original.

Low Latitude Glacial Deposits

Sedimentary rocks that are deposited by glaciers have distinctive features that enable their identification. Long before the advent of the *Snowball Earth* hypothesis many Neoproterozoic sediments had been interpreted as having a glacial origin, including some apparently at tropical latitudes at the time of their deposition. However, it is worth remembering that many sedimentary features traditionally associated with glaciers can also be formed by other means. Thus the glacial origin of many of the key occurrences for Snowball Earth has been contested. As of 2007, there is only one "very reliable" – still challenged – datum point identifying tropical tillites, which makes statements of equatorial ice cover somewhat presumptuous. Evidence of possible glacial origin of sediment includes:

- Dropstones (stones dropped into marine sediments), which can be deposited by glaciers or other phenomena.
- Varves (annual sediment layers in periglacial lakes), which can form at higher temperatures.
- Glacial striations (formed by embedded rocks scraped against bedrock): similar striations are from time to time formed by mudflows or tectonic movements.
- Diamictites (poorly sorted conglomerates). Originally described as glacial till, most were in fact formed by debris flows.

Open-Water Deposits

It appears that some deposits formed during the Snowball period could only have been formed in the presence of an active hydrological cycle. Bands of glacial deposits up to hundreds of meters thick, separated by small (meters) bands of non-glacial sediments, demonstrate that glaciers were melting and re-forming repeatedly; solid oceans would not permit this scale of deposition. It is considered possible that ice streams such as seen in Antarctica today could be responsible for these sequences. Further, sedimentary features that could only form in open water, for example wave-formed ripples, far-traveled ice-rafted debris and indicators of photosynthetic activity, can be found throughout sediments dating from the 'Snowball Earth' periods. While these may represent 'oases' of meltwater on a completely frozen Earth, computer modelling suggests that large areas of the ocean must have remained ice-free arguing that a "hard" snowball is not plausible in terms of energy balance and general circulation models.

Carbon Isotope Ratios

There are two stable isotopes of carbon in sea water: carbon-12 (^{12}C) and the rare carbon-13 (^{13}C), which makes up about 1.109 percent of all carbon isotopes.

Biochemical processes, of which photosynthesis is one, tend to preferentially incorporate the lighter ^{12}C isotope. Thus ocean-dwelling photosynthesizers, both protists and algae, tend to be very slightly depleted in ^{13}C, relative to the abundance found in the primary volcanic sources of the Earth's carbon. Therefore, an ocean with photosynthetic life will have a higher $^{12}C/^{13}C$ ratio within organic remains, and a lower ratio in corresponding ocean water. The organic component of the lithified sediments will forever remain very slightly, but measurably, depleted in ^{13}C.

During the proposed episode of Snowball Earth, there are rapid and extreme negative excursions in the ratio of ^{13}C to ^{12}C. This is consistent with a deep freeze that killed off most or nearly all photosynthetic life – although other mechanisms, such as clathrate release, can also cause such perturbations. Close analysis of the timing of ^{13}C 'spikes' in deposits across the globe allows the recognition of four, possibly five, glacial events in the late Neoproterozoic. However, the stratigraphic record of Oman presents a large negative carbon isotope excursion (within the Shuram Formation) away from any glacial evidence.

Banded Iron Formations

Banded iron formations (BIF) are sedimentary rocks of layered iron oxide and iron-poor chert. In the presence of oxygen, iron naturally rusts and becomes insoluble in water. The banded iron formations are commonly very old and their deposition is often related to the oxidation of the Earth's atmosphere during the Paleoproterozoic era, when dissolved iron in the ocean came in contact with photosynthetically-produced oxygen and precipitated out as iron oxide. The bands were produced at the tipping point between an anoxic and an oxygenated ocean. Since today's atmosphere is oxygen rich (nearly 21 percent by volume) and in contact with the oceans, it is not possible to accumulate enough iron oxide to deposit a banded formation. The only extensive iron formations that were deposited after the Paleoproterozoic (after 1.8 billion years ago) are associated with Cryogenian glacial deposits.

For such iron-rich rocks to be deposited there would have to be anoxia in the ocean, so that much dissolved iron (as ferrous oxide) could accumulate before it met an oxidant that would precipitate it as ferric oxide. For the ocean to become anoxic it must have limited gas exchange with the oxygenated atmosphere. Proponents of the hypothesis argue that the reappearance of BIF in the sedimentary record is a result of limited oxygen levels in an ocean sealed by sea ice, while opponents suggest that the rarity of the BIF deposits may indicate that they formed in inland seas. Being isolated from the oceans, such lakes may have been stagnant and anoxic at depth, much like today's Black Sea; a sufficient input of iron could provide the necessary conditions for BIF formation. A further difficulty in suggesting that BIFs marked the end of the glaciation is that they are found interbedded with glacial sediments. BIFs are also strikingly absent during the Marinoan glaciation.

Cap Carbonate Rocks

Around the top of Neoproterozoic glacial deposits there is commonly a sharp transition into a chemically precipatated sedimentary limestone or dolostone metres to tens of metres thick. These cap carbonates sometimes occur in sedimentary successions that have no other carbonate rocks, suggesting that their deposition is result of a profound aberration in ocean chemistry.

These cap carbonates have unusual chemical composition, as well as strange sedimentary structures that are often interpreted as large ripples. The formation of such sedimentary rocks could be caused by a large influx of positively-charged ions, as would be produced by rapid weathering during the extreme greenhouse following a Snowball Earth event. The $\delta^{13}C$ isotopic signature of the cap carbonates is near -5‰, consistent with the value of the mantle — such a low value is usually/could be taken to signify an absence of life, since photosynthesis usually acts to raise the value; alternatively the release of methane deposits could have lowered it from a higher value, and counterbalance the effects of photosynthesis.

The precise mechanism involved in the formation of cap carbonates is not clear, but the most cited explanation suggests that at the melting of a Snowball Earth, water would dissolve the abundant CO_2 from the atmosphere to form carbonic acid, which would fall as acid rain. This would weather exposed silicate and carbonate rock (including readily-attacked glacial debris), releasing large amounts of calcium, which when washed into the ocean would form distinctively textured layers of carbonate sedimentary rock. Such an abiotic "cap carbonate" sediment can be found on top of the glacial till that gave rise to the Snowball Earth hypothesis.

However, there are some problems with the designation of a glacial origin to cap carbonates. Firstly, the high carbon dioxide concentration in the atmosphere would cause the oceans to become acidic, and dissolve any carbonates contained within — starkly at odds with the deposition of cap carbonates. Further, the thickness of some cap carbonates is far above what could reasonably be produced in the relatively quick deglaciations. The cause is further weakened by the lack of cap carbonates above many sequences of clear glacial origin at a similar time and the occurrence of similar carbonates within the sequences of proposed glacial origin. An alternative mechanism, which may have produced the Doushantuo cap carbonate at least, is the rapid, widespread release of methane. This accounts for incredibly low - as low as 48‰ - $\delta^{13}C$ values — as well as unusual sedimentary features which appear to have been formed by the flow of gas through the sediments.

Changing Acidity

Isotopes of the element boron suggest that the pH of the oceans dropped dramatically before and after the Marinoan glaciation. This may indicate a build up of carbon dioxide in the atmosphere, some of which would dissolve into the oceans to form carbonic acid. Although the boron variations may be evidence of extreme climate change, they need not imply a global glaciation.

Space Dust

The Earth's surface is very depleted in the element iridium, which primarily resides in the Earth's core. The only significant source of the element at the surface is cosmic particles that reach Earth. During a Snowball Earth, iridium would accumulate on the ice sheets, and when the ice melted the resulting layer of sediment would be rich in iridium. An iridium anomaly has been discovered at the base of the cap carbonate formations, and has been used to suggest that the glacial episode lasted for at least 3 million years, but this does not necessarily imply a *global* extent to the glaciation; indeed a similar anomaly could be explained by the impact of a large extra-planetary object, such as a meteor.

Cyclic Climate Fluctuations

Using the ratio of mobile cations to those that remain in soils during chemical weathering (the chemical index of alteration), it has been shown that chemical weathering varied in a cyclic fashion within a glacial succession, increasing during interglacial periods and decreasing during cold and arid glacial periods. This pattern, if a true reflection of events, suggests that the "snowball Earths" bore a stronger resemblance to Pleistocene ice age cycles than to a completely frozen Earth.

What's more, glacial sediments of the Portaskaig formation in Scotland clearly show interbedded cycles of glacial and shallow marine sediments. The significance of these deposits is highly reliant upon their dating. Glacial sediments are difficult to date, and the closest dated bed to the Portaskaig group is 8 km stratigraphically above the beds of interest. Its dating to 600Ma means the beds can be tentatively correlated to the Sturtian glaciation, but they may represent the advance or retreat of a Snowball Earth.

Mechanisms

The initiation of a Snowball Earth event would involve some initial cooling mechanism, which would result in an increase in the Earth's coverage of snow and ice. The increase in Earth's coverage of snow and ice would increase the Earth's albedo, which would result in positive feedback for cooling. If enough snow and ice accumulates, runaway cooling would result. This positive feedback is facilitated by an equatorial continental distribution, which would allow ice to accumulate in the regions closer to the equator, where solar radiation is most direct.

The eruption of a super volcano could act as a trigger. A reduction in the atmospheric concentration of greenhouse gases such as Methane and/or Carbon dioxide could also act as a trigger. Alternatively, changes in solar energy output or perturbations of the Earth's orbit could also act as a trigger. Regardless of the trigger, initial cooling results in an increase in the area of the Earth's surface covered by snow and ice, and the additional snow and ice reflects more solar energy back to space, further cooling the Earth and further increasing the area of the Earth's surface covered by snow and ice. This positive feedback loop could eventually produce a frozen equator as cold as modern-day Antarctica.

Global warming associated with large accumulations of Carbon dioxide in the atmosphere over millions of years, emitted primarily by volcanic activity, is the proposed trigger for melting a Snowball Earth. Due to positive feedback for melting, the eventual melting of the snow and ice covering most of the Earth's surface would require as few as 1,000 years.

Modeling Disputes

While the presence of glaciers is not disputed, the idea that the entire planet was covered in ice is more contentious, leading some scientists to posit a "slushball Earth", in which a band of ice-free, or ice-thin, waters remains around the equator, allowing for a continued hydrologic cycle. This theory appeals to scientists who observe certain features of the sedimentary record that can only be formed under open water, or rapidly moving ice (which would require somewhere ice-free to move to). Recent research observed geochemical cyclicity in clastic rocks, showing that the "Snowball" periods were punctuated by warm spells, similar to ice age cycles in recent Earth history. Attempts to construct computer models of a Snowball Earth have also struggled to accommodate global ice cover without fundamental changes in the laws and constants which govern the planet.

Initiating "Snowball Earth"

A tropical distribution of the continents is, perhaps counter-intuitively, necessary to allow the initiation of a Snowball Earth. Firstly, tropical continents are more reflective than open ocean, and so absorb less of the sun's heat: most absorption of solar energy on Earth today occurs in tropical oceans.

Further, tropical continents are subject to more rainfall, which leads to increased river discharge — and erosion. When exposed to air, silicate rocks undergo weathering reactions which remove carbon dioxide from the atmosphere. These reactions proceed in the general form: Rock-forming mineral + CO_2 + H_2O '! cations + bicarbonate + SiO_2. An example of such a reaction is the weathering of wollastonite:

$$CaSiO_3 + 2CO_2 + H_2O \text{ '! } Ca^{2+} + SiO_2 + 2HCO_3^-$$

The released calcium cations react with the dissolved bicarbonate in the ocean to form calcium carbonate as a chemically precipitated sedimentary rock. This transfers carbon dioxide, a greenhouse gas, from the air into the geosphere, and, in steady-state on geologic time scales, offsets the carbon dioxide emitted from volcanoes into the atmosphere.

A paucity of suitable sediments for analysis makes precise continental distribution during the Neoproterozoic difficult to establish. Some reconstructions point towards polar continents — which have been a feature of all other major glaciations, providing a point upon which ice can nucleate.

Changes in ocean circulation patterns may then have provided the trigger of snowball Earth.

Additional factors that may have contributed to the onset of the Neoproterozoic Snowball include the introduction of atmospheric free oxygen, which may have reached sufficient quantities to react with methane in the atmosphere, oxidizing it to carbon dioxide, a much weaker greenhouse gas, and a younger — thus fainter — sun, which would have emitted 6 percent less radiation in the Neoproterozoic.

Normally, as the Earth gets colder due to natural climatic fluctuations and changes in incoming solar radiation, the cooling slows these weathering reactions. As a result, less carbon dioxide is removed from the atmosphere and the Earth warms as this greenhouse gas accumulates — this 'negative feedback' process limits the magnitude of cooling. During the Cryogenian period, however, the Earth's continents were all at tropical latitudes, which made this moderating process less effective, as high weathering rates continued on land even as the Earth cooled. This let ice advance beyond the polar regions. Once ice advanced to within 30° of the equator, a positive feedback could ensue such that the increased reflectiveness (albedo) of the ice led to further cooling and the formation of more ice, until the whole Earth is ice covered.

Polar continents, due to low rates of evaporation, are too dry to allow substantial carbon deposition — restricting the amount of atmospheric carbon dioxide that can be removed from the carbon cycle. A gradual rise of the proportion of the isotope carbon-13 relative to carbon-12 in sediments pre-dating "global" glaciation indicates that CO_2 draw-down before snowball Earths was a slow and continuous process.

The start of Snowball Earths are always marked by a sharp downturn in the $\delta^{13}C$ value of sediments, a hallmark that may be attributed to a crash in biological productivity as a result of the cold temperatures and ice-covered oceans.

During the Frozen Period

Global temperature fell so low that the equator was as cold as modern-day Antarctica. This low temperature was maintained by the reflective ice, its high albedo resulting in most incoming solar energy being reflected into space. A lack of heat-retaining clouds, caused by water vapor freezing out of the atmosphere, amplified this effect.

Breaking Out of Global Glaciation

The carbon dioxide levels necessary to unfreeze the Earth have been estimated as being 350 times what they are today, about 13% of the atmosphere. Since the Earth was almost completely covered with ice, carbon dioxide could not be withdrawn from the atmosphere by release of alkaline metal ions weathering out of siliceous rocks. Over 4 to 30 million years, enough CO_2 and methane, mainly emitted by volcanoes, would accumulate to finally cause enough greenhouse effect to make surface ice melt in the tropics until a band of permanently ice-free land and water developed; this would be darker than the ice, and thus absorb more energy from the sun — initiating a "positive feedback."

On the continents, the melting of glaciers would release massive amounts of glacial deposit, which would erode and weather. The resulting sediments supplied to the ocean would be high in nutrients such as phosphorus, which combined with the abundance of CO_2 would trigger a cyanobacteria population explosion, which would cause a relatively rapid reoxygenation of the atmosphere, which may have contributed to the rise of the Ediacaran biota and the subsequent Cambrian explosion — a higher oxygen concentration allowing large multicellular lifeforms to develop. This positive feedback loop would melt the ice in geological short order, perhaps less than 1,000 years; replenishment of atmospheric oxygen and depletion of the CO_2 levels would take further millennia.

Destabilization of substantial deposits of methane hydrates locked up in low-latitude permafrost may also have acted as a trigger and/or strong positive feedback for deglaciation and warming.

It is possible that carbon dioxide levels fell enough for Earth to freeze again; this cycle may have repeated until the continents had drifted to more polar latitudes.

Scientific Dispute

The leading argument against the hypothesis is evidence of fluctuation in ice cover and melting during "Snowball Earth" deposits. Evidence for such melting comes from evidence of glacial dropstones, geochemical evidence of climate cyclicity, and interbedded glacial and shallow marine sediments. A longer record from Oman, constrained to 13°N, covers the period from 712 to 545 million years ago — a time span containing the Sturtian and Marinoan glaciations — and shows both glacial and ice-free deposition.

There have been difficulties in recreating a snowball Earth with global climate models. Simple GCMs with mixed-layer oceans can be made to freeze to the equator; a more sophisticated model with a full dynamic ocean (though only a primitive sea ice model) failed to form sea ice to the equator . In addition, the levels of CO_2 necessary to melt a global ice cover have been calculated to be 130,000 ppm, which is considered by some to be unreasonably large.

Strontium isotopic data have been found to be at odds with proposed snowball Earth models of silicate weathering shutdown during glaciation and rapid rates immediately post-glaciation. Therefore, methane release from permafrost during marine transgression was proposed to be the source of the large measured carbon excursion in the time immediately after glaciation.

"Zipper Rift" Hypothesis

Some scholars suggest that the Neoproterozoic Snowball Earth was in fact no different from any other glaciation in Earth's history, and that efforts to find a single cause are likely to end in failure. The "Zipper rift" hypothesis proposes two pulses of continental "unzipping"—first, the breakup of the supercontinent Rodinia, forming the proto-Pacific ocean; then the splitting of the continent Baltica from Laurentia, forming the proto-Atlantic—coincided with the glaciated periods. The associated tectonic uplift would form high plateaus, just as the East African rift is responsible for high topography; this high ground could then host glaciers. Banded iron formations have been taken as unavoidable evidence for global ice cover, since they require dissolved iron ions and anoxic waters to form; however, the limited extent of the Neoproterozoic banded iron deposits means that they may not have formed in frozen oceans, but instead in inland seas. Such seas can experience a wide range of chemistries; high rates of evaporation could concentrate iron ions, and a periodic lack of circulation could allow anoxic bottom water to form. Continental rifting, with associated subsidence, tends to produce such landlocked water bodies. This rifting, and associated subsidence, would produce the space for the fast deposition of sediments, negating the need for an immense and rapid melting to raise the global sea levels.

High-Obliquity Hypothesis

A competing theory to explain the presence of ice on the equatorial continents was that the Earth's axial tilt was quite high, in the vicinity of 60°, which would place the Earth's land in high "latitudes", although supporting evidence is scarce. A less extreme possibility would be that it was merely the Earth's magnetic pole that wandered to this inclination, as the magnetic readings which suggested ice-filled continents depends on the magnetic and rotational poles being relatively similar. In either of these two situations, the freeze would be limited to relatively small areas, as is the case today; severe changes to the Earth's climate are not necessary.

Inertial Interchange True Polar Wander

The evidence for low latitude glacial deposits during the supposed Snowball Earth episodes has been reinterpreted via the concept of inertial interchange true polar wander (IITPW). This theory, created to explain palaeomagnetic data, suggests that the continents drifted far faster during the late Neoproterozoic, allowing glacial deposits to form at the poles before continents returned to the equator, when palaeomagnetic beds were laid down. While the physics behind the proposition is sound, the removal of one flawed data point from the original study rendered the application of the concept in these circumstances unwarranted.

Several alternative explanations for the evidence have been proposed.

Survival of Life through Frozen Periods

A tremendous glaciation would curtail plant life on Earth, thus letting the atmospheric oxygen be drastically depleted and perhaps even disappear, and thus allow non-oxidized iron-rich rocks to form.

Detractors argue that this kind of glaciation would have made life extinct entirely. However, microfossils such as stromatolites and oncolites prove that in shallow marine environments at least life did not suffer any perturbation. Instead life developed a trophic complexity and survived the cold period unscathed. Proponents counter that it may have been possible for life to survive in these ways:

- Reservoirs of anaerobic and low-oxygen life powered by chemicals in deep oceanic hydrothermal vents surviving in Earth's deep oceans and crust; but photosynthesis would not have been possible there.

- As eggs and dormant cells and spores deep-frozen into ice right through the worst phases of the frozen period.
- Under the ice layer, in chemolithotrophic (mineral-metabolizing) ecosystems theoretically resembling those in existence in modern glacier beds, high-alpine and Arctic talus permafrost, and basal glacial ice. This is especially plausible in areas of volcanism or geothermal activity.
- In deep ocean regions far from the supercontinent Rodinia or its remnants as it broke apart and drifted on the tectonic plates, which may have allowed for some small regions of open water preserving small quantities of life with access to light and CO_2 for photosynthesizers (not multicellular plants, which did not yet exist) to generate traces of oxygen that were enough to sustain some oxygen-dependent organisms. This would happen even if the sea froze over completely if small parts of the ice were thin enough to admit light.
- In nunatak areas in the tropics, where daytime tropical sun or volcanic heat heated bare rock sheltered from cold wind and made small temporary melt pools, which would freeze over at sunset.
- In pockets of liquid water within and under the ice caps, similar to Lake Vostok in Antarctica. In theory, this system may resemble microbial communities living in the perennially frozen lakes of the Antarctic dry valleys. Photosynthesis can occur under ice up to 100 m thick, and at the temperatures predicted by models equatorial sublimation would prevent equatorial ice thickness from exceeding 10 m.

- In small oases of liquid water, as would be found near geothermal hotspots resembling Iceland today.

However, organisms and ecosystems, as far as it can be determined by the fossil record, do not appear to have undergone the significant change that would be expected by a mass extinction. With the advent of more precise dating, a phytoplankton extinction event which had been associated with Snowball Earth was shown to precede glaciations by 16 million years. Even if life were to cling on in all the ecological refuges listed above, a whole-Earth glaciation would result in a biota with a noticeably different diversity and composition. This change in diversity and composition has not yet been observed– in fact, the organisms which ought to be most susceptible to climatic variation emerge unscathed from the Snowball Earth.

Implications

A Snowball Earth has profound implications in the history of life on Earth. While many refugia have been postulated, global ice cover would certainly have ravaged ecosystems dependent on sunlight. Geochemical evidence from rocks associated with low-latitude glacial deposits have been interpreted to show a crash in oceanic life during the glacials.

The melting of the ice may have presented many new opportunities for diversification, and may indeed have driven the rapid evolution which took place at the end of the Cryogenian period.

Effect on Early Evolution

The Neoproterozoic was a time of remarkable diversification of multicellular organisms, including animals. Organism size and complexity increased considerably after the end of the Snowball glaciations. This development of multicellular organisms may have been the result of increased evolutionary pressures resulting from multiple icehouse-hothouse cycles; in this sense, Snowball Earth episodes may have "pumped" evolution. Alternatively, fluctuating nutrient levels and rising oxygen may have played a part. Interestingly, another major glacial episode may have ended just a few million years before the Cambrian explosion.

Mechanistically, the impact of snowball Earth (in particular the later glaciations) on complex life is likely to have occurred through the process of kin selection. Organ-scale differentiation, in particular the terminal (irreversible) differentiation present in animals, requires the individual cell (and the genes contained within it) to "sacrifice" their ability to reproduce, so that the colony is not disrupted. From the short-term perspective of the gene, more offspring will be gained (in the short-term) by causing the cell in which it is contained to ignore any signals received from the colony, and to reproduce at the maximum rate, regardless of the implications for the wider group. Today, this incentive explains the formation of tumours in animals and plants.

Such costly, "altruistic" differentiation can be adaptive (maximise the number of surviving offspring) to individual genes if the consequence of altruism (terminal cellular differentiation) benefits other copies of such genes. (Note that "altruism" refers only to the reproductive cost of the trait, and implies no sentience or foresight). Because relatives share genes, genes causing altruism (such as organ scale differentiation) can spread if it occurs between relatives, see kin selection.

It has been argued that because snowball Earth would undoubtedly have decimated the population size of any given species, the extremely small populations that resulted would all have been descended from a small number of individuals (see founder effect), and consequently the average relatedness between any two individuals (in this case individual cells) would have been exceptionally high as a result of glaciations. Altruism is known to increase from rarity when relatedness (R) exceeds the ratio of the cost (C) to the altruist (in this case, the cell giving up its own reproduction by differentiating), to the benefit (B) to the recipient of altruism (the germ line of the colony, that reproduces as a result of the differentiation), i.e. R > C/B (see Hamilton's rule). The evolutionary pressure of the high relatedness caused by the glaciations may have been sufficient to overcome the reproductive cost of forming a complex animal, for the first time in Earth's history.

Occurrence and timing of Snowball Earths

Neoproterozoic

There are three or four significant ice ages during the late Neoproterozoic. Of these, the Marinoan was the most significant, and the Sturtian glaciations were also truly widespread. Even the leading Snowball proponent Hoffman agrees that the ~million year long Gaskiers glaciation did not lead to global glaciation, although it was probably as intense as the late Ordovician glaciation. The status of the Kaigas "glaciation" or "cooling event" is currently unclear; some workers do not recognise it as a glacial, others suspect that it may reflect poorly dated strata of Sturtian association, and others believe it may indeed be a third ice age. It was certainly less significant than the Sturtian or Marinoan glaciations, and probably not global in extent. Emerging evidence suggests that the Earth underwent a number of glaciations during the Neoproterozoic, which would stand strongly at odds with the Snowball hypothesis.

Paleoproterozoic

The Snowball Earth hypothesis has been invoked to explain glacial deposits in the Huronian supergroup of Canada though the palaeomagnetic evidence that suggests ice sheets at low latitudes is contested. The glacial sediments of the Makganyene formation of South Africa are slightly younger than the Huronian glacial deposits (~2.25 billion years old) and were deposited at tropical latitudes. It has been proposed that rise of free oxygen that occurred during this part of the Paleoproterozoic removed methane in the atmosphere through oxidation. As the Sun was notably weaker at the time, the Earth's climate may have relied on methane, a powerful greenhouse gas, to maintain surface temperatures above freezing. In the absence of this methane greenhouse, temperatures plunged and a snowball event could have occurred.

Karoo Ice Age

Before the theory of continental drift, glacial deposits in Carboniferous strata in tropical continents areas such as India and South America led to speculation that the Karoo Ice Age glaciation reached into the tropics. However, a continental reconstruction shows that ice was in fact constrained to the polar parts of the supercontinent Gondwana.

Current Field Work

Antarctica

Stephen Warren, a professor from the University of Washington in Seattle, will lead a small team of U.S. researchers to Antarctica for the 2009-10 and 2010-11 summer field seasons to add more data in support of the Snowball Earth hypothesis. A project funded by the National Science Foundation, the project will study different ice types found in Antarctica that may have existed during the previous Snowball Earth events, specifically to learn more about their albedo, or reflectivity, information important in understanding the processes involved in the exteme glaciation. The team will spend time on the sea ice in McMurdo Sound and in Garwood Valley in the McMurdo Dry Valleys the first season. The second season will be spent in the Transantarctic Mountains.

6
Climate and Changes

Climate encompasses the statistics of temperature, humidity, atmospheric pressure, wind, rainfall, atmospheric particle count and numerous other meteorological elements in a given region over long periods of time. Climate can be contrasted to weather, which is the present condition of these same elements over periods up to two weeks.

The climate of a location is affected by its latitude, terrain, altitude, ice or snow cover, as well as nearby water bodies and their currents. Climates can be classified according to the average and typical ranges of different variables, most commonly temperature and rainfall. The most commonly used classification scheme is the one originally developed by Wladimir Köppen. The Thornthwaite system, in use since 1948, incorporates evapotranspiration in addition to temperature and precipitation information and is used in studying animal species diversity and potential impacts of climate changes. The Bergeron and Spatial Synoptic Classification systems focus on the origin of air masses defining the climate for certain areas.

Paleoclimatology is the study and description of ancient climates. Since direct observations of climate are not available before the 19th century, paleoclimates are inferred from *proxy variables* that include non-biotic evidence such as sediments found in lake beds and ice cores, and biotic evidence such as tree rings and coral. Climate models are mathematical models of past, present and future climates.

Definition

Climate (from Ancient Greek *klima*, meaning *inclination*) is commonly defined as the weather averaged over a long period of time. The standard averaging period is 30 years, but other periods may be used depending on the purpose. Climate also includes statistics other than the average, such as the magnitudes of day-to-day or year-to-year variations. The Intergovernmental Panel on Climate Change (IPCC) glossary definition is:

Climate in a narrow sense is usually defined as the "average weather," or more rigorously, as the statistical description in terms of the mean and variability of relevant quantities over a period of time ranging from months to thousands or millions of years. The classical period is 30 years, as defined by the World Meteorological Organization (WMO). These quantities are most often surface variables such as temperature, precipitation, and wind. Climate in a wider sense is the state, including a statistical description, of the climate system.

The difference between climate and weather is usefully summarized by the popular phrase "Climate is what you expect, weather is what you get." Over historical time spans there are a number of nearly constant variables that determine climate, including latitude, altitude, proportion of land to water, and proximity to oceans and mountains. These change only over periods of millions of years due to processes such as plate tectonics. Other climate determinants are more dynamic: for example, the thermohaline circulation of the ocean leads to a 5°C (9°F) warming of the northern Atlantic ocean compared to other ocean basins. Other ocean currents redistribute heat between land and water on a more regional scale. The density and type of vegetation coverage affects solar heat absorption, water retention, and rainfall on a regional level. Alterations in the quantity of atmospheric greenhouse gases determines the amount of solar energy retained by the planet, leading to global warming or global cooling. The variables which determine climate are numerous and the interactions complex, but there is general agreement that the broad outlines are understood, at least insofar as the determinants of historical climate change are concerned.

Climate Classification

There are several ways to classify climates into similar regimes. Originally, climes were defined in Ancient Greece to describe the weather depending upon a location's latitude. Modern climate classification methods can be broadly divided into *genetic* methods, which focus on the causes of climate, and *empiric* methods, which focus on the effects of climate. Examples of genetic classification include methods based on the relative frequency of different air mass types or locations within synoptic weather disturbances. Examples of empiric classifications include climate zones defined by plant hardiness, evapotranspiration, or more generally the Köppen climate classification which was originally designed to identify the climates associated with certain biomes. A common shortcoming of these classification schemes is that they produce distinct boundaries between the zones they define, rather than the gradual transition of climate properties more common in nature.

Bergeron and Spatial Synoptic

The most generic classification is that involving the concept of air masses. The Bergeron classification is the most widely accepted form of air mass classification. Air mass classification involves three letters. The first letter describes its moisture properties, with c used for continental air masses (dry) and m for maritime air masses (moist). The second letter describes the thermal characteristic of its source region: T for tropical, P for polar, A for Arctic or Antarctic, M for monsoon,

E for equatorial, and S for superior air (dry air formed by significant downward motion in the atmosphere). The third letter is used to designate the stability of the atmosphere. If the air mass is colder than the ground below it, it is labelled k. If the air mass is warmer than the ground below it, it is labelled w. While air mass identification was originally used in weather forecasting during the 1950s, climatologists began to establish synoptic climatologies based on this idea in 1973.

Based upon the Bergeron classification scheme is the Spatial Synoptic Classification system (SSC). There are six categories within the SSC scheme: Dry Polar (similar to continental polar), Dry Moderate (similar to maritime superior), Dry Tropical (similar to continental tropical), Moist Polar (similar to maritime polar), Moist Moderate (a hybrid between maritime polar and maritime tropical), and Moist Tropical (similar to maritime tropical, maritime monsoon, or maritime equatorial).

Koppen

The Koppen classification depends on average monthly values of temperature and precipitation. The most commonly used form of the Koppen classification has five primary types labelled A through E. Specifically, the primary types are A, tropical; B, dry; C, mild mid-latitude; D, cold mid-latitude; and E, polar. The five primary classifications can be further divided into secondary classifications such as rain forest, monsoon, tropical savanna, humid subtropical, humid continental, oceanic climate, Mediterranean climate, steppe, subarctic climate, tundra, polar ice cap, and desert.

Rain forests are characterized by high rainfall, with definitions setting minimum normal annual rainfall between 1,750 millimetres (69 in) and 2,000 millimetres (79 in). Mean monthly temperatures exceed 18°C (64°F) during all months of the year.

A monsoon is a seasonal prevailing wind which lasts for several months, ushering in a region's rainy season. Regions within North America, South America. Sub-Saharan Africa, Australia and East Asia are monsoon regimes.

A tropical savanna is a grassland biome located in semi-arid to semi-humid climate regions of subtropical and tropical latitudes, with average temperatures remain at or above 18 °C (64°F) year round and rainfall between 750 millimetres (30 in) and 1,270 millimetres (50 in) a year. They are widespread on Africa, and are also found in India, the northern parts of South America, Malaysia, and Australia.

The humid subtropical climate zone where winter rainfall (and sometimes snowfall) is associated with large storms that the westerlies steer from west to east. Most summer rainfall occurs during thunderstorms and from occasional tropical cyclones. Humid subtropical climates lie on the east side continents, roughly between latitudes 20° and 40° degrees away from the equator.

A humid continental climate is marked by variable weather patterns and a large seasonal temperature variance. Places with a hottest monthly temperature above 10°C (50°F) and a coldest month temperature below "3°C (26.6°F) and which do not meet the criteria for an arid climate, are classified as continental.

An oceanic climate is typically found along the west coasts at the middle latitudes of all the world's continents, and in southeastern Australia, and is accompanied by plentiful precipitation year round.

The Mediterranean climate regime resembles the climate of the lands in the Mediterranean Basin, parts of western North America, parts of Western and South Australia, in southwestern South Africa and in parts of central Chile. The climate is characterized by hot, dry summers and cool, wet winters.

A steppe is a dry grassland with an annual temperature range in the summer of up to 40°C (104°F) and during the winter down to "40°C ("40.0°F).

A subarctic climate has little precipitation, and monthly temperatures which are above 10°C (50°F) for one to three months of the year, with continuous permafrost due to the very cold winters. Winters within subarctic climates include up to six months of temperatures averaging below 0°C (32°F).

Tundra occurs in the far Northern Hemisphere, north of the taiga belt, including vast areas of northern Russia and Canada.

A polar ice cap, or polar ice sheet, is a high-latitude region of a planet or moon that is covered in ice. Ice caps form because high-latitude regions receive less energy in the form of solar radiation from the sun than equatorial regions, resulting in lower surface temperatures.

A desert is a landscape form or region that receives very little precipitation. Deserts usually have a large diurnal and seasonal temperature range, with high daytime temperatures (in summer up to 45°C or 113°F), and low night-time temperatures (in winter down to 0°C; 32°F) due to extremely low humidity. Many deserts are formed by rain shadows, as mountains block the path of moisture and precipitation to the desert.

Thornthwaite

Devised by the American climatologist and geographer C.W. Thornthwaite, this climate classification method monitors the soil water budget using the concept of evapotranspiration. It monitors the portion of total precipitation used to nourish vegetation over a certain area. It uses indices such as a humidity index and an aridity index to determine an area's moisture regime based upon its average temperature, average rainfall, and average vegetation type. The lower the value of the index is any given area, the drier the area is.

The moisture classification includes climatic classes with descriptors such as hyperhumid, humid, subhumid, subarid, semi-arid (values of -20 to -40), and arid (values below -40). Humid regions experience more precipitation than evaporation each year, while arid regions experience greater evaporation than precipitation on an annual basis. A total of 33 per cent of the Earth's landmass is considered either arid of semi-arid, including southwest North America, southwest South America, most of northern and a small part of southern Africa, southwest and portions of eastern Asia, as well as much of Australia. Studies suggest that precipitation effectiveness (PE) within the Thornthwaite moisture index is overestimated in the summer and underestimated in the winter. This index can be effectively used to determine the number of herbivore and mammal species numbers within a given area. The index is also used in studies of climate change.

Thermal classifications within the Thornthwaite scheme include microthermal, mesothermal, and megathermal regimes. A mircothermal climate is one of low annual mean temperatures, generally between 0°C (32°F) and 14°C (57°F) which experiences short summers and has a potential evaporation between 14 centimetres (5.5 in) and 43 centimetres (17 in). A mesothermal climate lacks persistent heat or persistent cold, with potential evaporation between 57 centimetres (22 in) and 114 centimetres (45 in). A megathermal climate is one with persistent high temperatures and abundant rainfall, with potential evaporation in excess of 114 centimetres (45 in).

Record

Modern

Details of the modern climate record are known through the taking of measurements from such weather instruments as thermometers, barometers, and anemometers during the past few centuries. The instruments used to study weather conditions over the modern time scale, their known error, their immediate environment, and their exposure have changed over the years, which must be considered when studying the climate of centuries past.

Paleoclimatology

Paleoclimatology is the study of past climate over a great period of the Earth's history. It uses evidence from ice sheets, tree rings, sediments, coral, and rocks to determine the past state of the climate. It demonstrates periods of stability and periods of change and can indicate whether changes follow patterns such as regular cycles.

Climate Change

Climate change refers to the variation in the Earth's global climate or in regional climates over time. It describes changes in the variability or average state of the atmosphere over time scales ranging from decades to millions of years. These changes can be caused by processes internal to the Earth, external forces (e.g. variations in sunlight intensity) or, more recently, human activities.

In recent usage, especially in the context of environmental policy, the term "climate change" often refers only to changes in modern climate, including the rise in average surface temperature known as global warming. In some cases, the term is also used with a presumption of human causation, as in the United Nations Framework Convention on Climate Change (UNFCCC). The UNFCCC uses "climate variability" for non-human caused variations.

Earth has undergone periodic climate shifts in the past, including four major ice ages. These consisting of glacial periods where conditions are colder than normal, separated by interglacial periods. The accumulation of snow and ice during a glacial period increases the surface albedo, reflecting more of the Sun's energy into space and maintaining a lower atmospheric temperature. Increases in greenhouse gases, such as by volcanic activity, can increase the global temperature and produce an interglacial. Suggested causes of ice age periods include the positions of the continents, variations in the Earth's orbit, changes in the solar output, and vulcanism.

Climate Models

Climate models use quantitative methods to simulate the interactions of the atmosphere, oceans, land surface and ice. They are used for a variety of purposes from study of the dynamics of the weather and climate system to projections of future climate. All climate models balance, or very nearly balance, incoming energy as short wave (including visible) electromagnetic radiation to the earth with outgoing energy as long wave (infrared) electromagnetic radiation from the earth. Any imbalance results in a change in the average temperature of the earth.

The most talked-about models of recent years have been those relating temperature to the build-up of greenhouse gases in the atmosphere, primarily carbon dioxide (see greenhouse gas). These models predict an upward trend in the global mean surface temperature, with the most rapid increase in temperature being projected for the higher latitudes of the Northern Hemisphere.

Models can range from relatively simple to quite complex:

- A simple radiant heat transfer model that treats the earth as a single point and averages outgoing energy
- this can be expanded vertically (radiative-convective models), or horizontally
- finally, (coupled) atmosphere–ocean–sea ice global climate models discretise and solve the full equations for mass and energy transfer and radiant exchange.

Climate Change

Climate change is any long-term change in the statistics of weather over periods of time that range from decades to millions of years. It can express itself as a change in the mean weather conditions, the probability of extreme conditions, or in any other part of the statistical distribution of weather. Climate change may occur in a specific region, or across the whole Earth.

In recent usage, especially in the context of environmental policy, climate change usually refers to changes in modern climate (see global warming). For information on temperature measurements over various periods, and the data sources available, see temperature record. For attribution of climate change over the past century, see attribution of recent climate change.

Causes

Factors that can shape climate are often called climate forcings. These include such processes as variations in solar radiation, deviations in the Earth's orbit, and changes in greenhouse gas concentrations. There are a variety of climate change feedbacks that can either amplify or diminish the initial forcing. Some parts of the climate system, such as the oceans and ice caps, respond slowly in reaction to climate forcing because of their large mass. Therefore, the climate system can take centuries or longer to fully respond to new external forcings.

Plate Tectonics

Over the course of millions of years, the motion of tectonic plates reconfigures global land and ocean areas and generates topography. This can affect both global and local patterns of climate and atmosphere-ocean circulation.

The position of the continents determines the geometry of the oceans and therefore influences patterns of ocean circulation. Because the circulation of the ocean and the atmosphere are fundamentally linked, the locations of the continents are important in controlling the transfer of heat and moisture across the globe, and therefore, in determining global climate. A recent example of tectonic control on ocean circulation is the formation of the Isthmus of Panama about 5 million years ago, which shut off direct mixing between the Atlantic and Pacific Oceans. This strengthened the Gulf Stream and eventually led to Northern Hemisphere ice cover. Earlier, during the Carboniferous period, plate tectonics may have triggered the large-scale storage of carbon and increased glaciation. Geologic evidence points to a "megamonsoonal" circulation pattern during the time of the supercontinent Pangaea, and climate modeling suggests that the existence of the supercontinent was conductive to the establishment of monsoons.

More locally, topography can influence climate. The existence of mountains (as a product of plate tectonics through mountain-building) can cause orographic precipitation. Humidity generally decreases and diurnal temperature swings generally increase with increasing elevation. Mean temperature and the length of the growing season also decrease with increasing elevation. This, along with orographic precipitation, is important for the existence of low-latitude alpine glaciers and the varied flora and fauna along at different elevations in montane ecosystems.

The size of continents is also important. Because of the stabilizing effect of the oceans on temperature, yearly temperature variations are generally lower in coastal areas than they are inland. A larger supercontinent will therefore have more area in which climate is strongly seasonal than will several smaller continents and/or island arcs.

Solar Output

The sun is the predominant source for energy input to the Earth. Both long- and short-term variations in solar intensity are noted to affect global climate.

Early in Earth's history the sun emitted only 70% as much power as it does today. With the same atmospheric composition as exists today, liquid water should not have existed on Earth. However, there is evidence for the presence of water on the early Earth, in the Hadean and Archean eons, leading to what is known as the faint young sun paradox. Hypothesized solutions to this paradox include a vastly different atmosphere, with much higher concentrations of greenhouse gases than currently exist, and a stronger solar wind that could shield the Earth from the cooling effects of cosmic rays. Over the following approximately 4 billion years, the energy output of the sun increased and atmospheric composition changed, with the oxygenation of the atmosphere being the most notable alteration. The luminosity of the sun will continue to increase as it follows the main sequence. These changes in luminosity, and the sun's ultimate death as it becomes a red giant and then a white dwarf, will have large effects on climate, with the red giant phase possibly ending life on Earth.

Solar output also varies on shorter time scales, including the 11-year solar cycle and longer-term modulations. The 11-year sunspot cycle produces only a small change in temperature near Earth's surface (on the order of a tenth of a degree) but has a greater influence in the atmosphere's upper layers. Solar intensity variations are considered to have been influential in triggering the Little Ice Age, and for some of the warming observed from 1900 to 1950. The cyclical nature of the sun's energy output is not yet fully understood; it differs from the very slow change that is happening within the sun as it ages and evolves, with some studies pointing toward solar radiation increases from cyclical sunspot activity affecting global warming

Orbital Variations

Slight variations in Earth's orbit lead to changes in the amount of sunlight reaching the Earth's surface and how it is distributed across the globe. The former is similar to solar variations in that there is a change to the power input from the sun to the Earth system. The latter is due to how the orbital variations affect when and where sunlight is received by the Earth. The three types of orbital variations are variations in Earth's eccentricity, changes in the tilt angle of Earth's axis of rotation, and precession of Earth's axis. Combined together, these produce Milankovitch cycles which have a large impact on climate and are notable for their correlation to glacial and interglacial periods, their correlation with the advance and retreat of the Sahara, and for their appearance in the stratigraphic record.

Volcanism

Volcanism is the process of conveying material from the crust and mantle of the Earth to its surface. Volcanic eruptions, geysers, and hot springs, are examples of volcanic processes which release gases and/or particulates into the atmosphere.

Eruptions large enough to affect climate occur on average several times per century, and cause cooling for a period of a few years. The eruption of Mount Pinatubo in 1991, the second largest terrestrial eruption of the 20th century (after the 1912 eruption of Novarupta) affected the climate substantially. Global temperatures decreased by about 0.5°C (0.9°F). Much larger eruptions, known as large igneous provinces, occur only a few times every hundred million years, but can reshape climate for millions of years and cause mass extinctions. Initially, it was thought that the dust ejected into the atmosphere from large volcanic eruptions was responsible for longer-term cooling by partially blocking the transmission of solar radiation to the Earth's surface. However, measurements indicate that most of the dust hurled into the atmosphere may return to the Earth's surface within as little as six months, given the right conditions.

Volcanoes are also part of the extended carbon cycle. Over very long (geological) time periods, they release carbon dioxide from the Earth's crust and mantle, counteracting the uptake by sedimentary rocks and other geological carbon dioxide sinks. According to the US Geological Survey, however, estimates are that human activities generate more than 130 times the amount of carbon dioxide emitted by volcanoes.

Ocean Variability

The ocean is a fundamental part of the climate system. Short-term fluctuations (years to a few decades) such as the El Niño–Southern Oscillation, the Pacific decadal oscillation, the North Atlantic oscillation, and the Arctic oscillation, represent climate variability rather than climate change. On longer time scales, alterations to ocean processes such as thermohaline circulation play a key role in redistributing heat by carrying out a very slow and extremely deep movement of water, and the long-term redistribution of heat in the world's oceans.

Human Influences

Anthropogenic factors are human activities that change the environment. In some cases the chain of causality of human influence on the climate is direct and unambiguous (for example, the effects of irrigation on local humidity), whilst in other instances it is less clear. Various hypotheses for human-induced climate change have been argued for many years. Presently the scientific consensus on climate change is that human activity is very likely the cause for the rapid increase in global average temperatures over the past several decades. Consequently, the debate has largely shifted onto ways to reduce further human impact and to find ways to adapt to change that has already occurred.

Of most concern in these anthropogenic factors is the increase in CO_2 levels due to emissions from fossil fuel combustion, followed by aerosols (particulate matter in the atmosphere) and cement manufacture. Other factors, including land use, ozone depletion, animal agriculture and deforestation, are also of concern in the roles they play — both separately and in conjunction with other factors — in affecting climate.

Physical Evidence for Climatic Change

Evidence for climatic change is taken from a variety of sources that can be used to reconstruct past climates. Reasonably complete global records of surface temperature are available beginning from the mid-late 1800s. For earlier periods, most of the evidence is indirect—climatic changes are inferred from changes in indicators that reflect climate, such as vegetation, ice cores, dendrochronology, sea level change, and glacial geology.

Historical and Archaeological Evidence

Climate change in the recent past may be detected by corresponding changes in settlement and agricultural patterns. Archaeological evidence, oral history and historical documents can offer insights into past changes in the climate. Climate change effects have been linked to the collapse of various civilisations.

Glaciers

Glaciers are among the most sensitive indicators of climate change, advancing when climate cools (for example, during the period known as the Little Ice Age) and retreating when climate warms. Glaciers grow and shrink, both contributing to natural variability and amplifying externally forced changes. A world glacier inventory has been compiled since the 1970s. Initially based mainly on aerial photographs and maps, this compilation has resulted in a detailed inventory of more than 100,000 glaciers covering a total area of approximately 240,000 km^2 and, in preliminary estimates, for the recording of the remaining ice cover estimated to be around 445,000 km^2. The World Glacier Monitoring Service collects data annually on glacier retreat and glacier mass balance From this data, glaciers worldwide have been found to be shrinking significantly, with strong glacier retreats in the 1940s, stable or growing conditions during the 1920s and 1970s, and again retreating from the mid 1980s to present. Mass balance data indicate 17 consecutive years of negative glacier mass balance.

The most significant climate processes since the middle to late Pliocene (approximately 3 million years ago) are the glacial and interglacial cycles. The present interglacial period (the Holocene) has lasted about 11,700 years. Shaped by orbital variations, responses such as the rise and fall of continental ice sheets and significant sea-level changes helped create the climate. Other changes, including Heinrich events, Dansgaard–Oeschger events and the Younger Dryas, however, illustrate how glacial variations may also influence climate without the forcing effect of orbital changes.

Glaciers leave behind moraines that contain a wealth of material — including organic matter that may be accurately dated — recording the periods in which a glacier advanced and retreated. Similarly, by tephrochronological techniques, the lack of glacier cover can be identified by the presence of soil or volcanic tephra horizons whose date of deposit may also be precisely ascertained.

Vegetation

A change in the type, distribution and coverage of vegetation may occur given a change in the climate; this much is obvious. In any given scenario, a mild change in climate may result in increased precipitation and warmth, resulting in improved plant growth and the subsequent sequestration of airborne CO_2. Larger, faster or more radical changes, however, may well result in vegetation stress, rapid plant loss and desertification in certain circumstances.

Ice Cores

Analysis of ice in a core drilled from a ice sheet such as the Antarctic ice sheet, can be used to show a link between temperature and global sea level variations. The air trapped in bubbles in the ice can also reveal the CO_2 variations of the atmosphere from the distant past, well before modern environmental influences. The study of these ice cores has been a significant indicator of the changes in CO_2 over many millennia, and continue to provide valuable information about the differences between ancient and modern atmospheric conditions.

Dendrochronology

Dendochronology is the analysis of tree ring growth patterns to determine the age of a tree. From a climate change viewpoint, however, Dendochronology can also indicate the climatic conditions for a given number of years. Wide and thick rings indicate a fertile, well-watered growing period, whilst thin, narrow rings indicate a time of lower rainfall and less-than-ideal growing conditions.

Pollen Analysis

Palynology is the study of contemporary and fossil palynomorphs, including pollen. Palynology is used to infer the geographical distribution of plant species, which vary under different climate conditions. Different groups of plants have pollen with distinctive shapes and surface textures, and since the outer surface of pollen is composed of a very resilient material, they resist decay. Changes in the type of pollen found in different sedimentation levels in lakes, bogs or river deltas indicate changes in plant communities; which are dependent on climate conditions.

Insects

Remains of beetles are common in freshwater and land sediments. Different species of beetles tend to be found under different climatic conditions. Given the extensive lineage of beetles whose genetic makeup has not altered significantly over the millennia, knowledge of the present climatic range of the different species, and the age of the sediments in which remains are found, past climatic conditions may be inferred.

Sea Level Change

Global sea level change for much of the last century has generally been estimated using tide gauge measurements collated over long periods of time to give a long-term average. More recently, altimeter measurements—in combination with accurately determined satellite orbits—have provided an improved measurement of global sea level change.

7
Atmosphere of Earth

The Earth's atmosphere is a layer of gases surrounding the planet Earth that is retained by the Earth's gravity. It has a mass of about five quadrillion metric tons. Dry air contains roughly (by volume) 78.08% nitrogen, 20.95% oxygen, 0.93% argon, 0.038% carbon dioxide, and trace amounts of other gases. Air also contains a variable amount of water vapor, on average around 1%. The atmosphere protects life on Earth by absorbing ultraviolet solar radiation, warming the surface through heat retention (greenhouse effect), and reducing temperature extremes between day and night.

There is no definite boundary between the atmosphere and outer space. It slowly becomes thinner and fades into space. An altitude of 120 km (75 mi) marks the boundary where atmospheric effects become noticeable during atmospheric reentry. The Kármán line, at 100 km (62 mi), is also frequently regarded as the boundary between atmosphere and outer space. Three quarters of the atmosphere's mass is within 11 km (6.8 mi; 36,000 ft) of the surface.

Temperature and Layers

The temperature of the Earth's atmosphere varies with altitude; the mathematical relationship between temperature and altitude varies among five different atmospheric layers (ordered highest to lowest, the ionosphere is part of the thermosphere):

Exosphere

From 500–1,000 km (310–620 mi; 1,600,000–3,300,000 ft) up to 10,000 km (6,200 mi; 33,000,000 ft), contain free-moving particles that may migrate into and out of the magnetosphere or the solar wind.

Exobase

Also known as the 'critical level', it is the lower boundary of the exosphere.

Ionosphere

The part of the atmosphere that is ionized by solar radiation stretches from 50 to 1,000 km (31 to 620 mi; 160,000 to 3,300,000 ft) and typically overlaps both the exosphere and the thermosphere. It plays an important part in atmospheric electricity and forms the inner edge of the magnetosphere. Because of its charged particles, it has practical importance because it influences, for example, radio propagation on the Earth. It is responsible for auroras.

Thermopause

The boundary above the thermosphere, it varies in height from 500–1,000 km (310–620 mi; 1,600,000–3,300,000 ft).

Thermosphere

From 80–85 km (50–53 mi; 260,000–280,000 ft) to over 640 km (400 mi; 2,100,000 ft), temperature increasing with height. The temperature of this layer can rise to 1,500°C (2,730°F). The International Space Station orbits in this layer, between 320 and 380 km (200 and 240 mi).

Mesopause

The temperature minimum at the boundary between the thermosphere and the mesosphere. It is the coldest place on Earth, with a temperature of "100°C ("148.0°F; 173.1 K).

Mesosphere

From the Greek word "ìÝóïò" meaning middle. The mesosphere extends from about 50 km (31 mi; 160,000 ft) to the range of 80–85 km (50–53 mi; 260,000–280,000 ft). Temperature decreases with height, reaching "100°C ("148.0°F; 173.1 K) in the upper mesosphere. This is also where most meteors burn up when entering the atmosphere.

Stratopause

The boundary between the mesosphere and the stratosphere, typically 50 to 55 km (31 to 34 mi; 160,000 to 180,000 ft). The pressure here is 1/1000th sea level.

Stratosphere

From the Latin word "stratus" meaning spreading out. The stratosphere extends from the troposphere's 7–17 km (4.3–11 mi; 23,000–56,000 ft) range to about 51 km (32 mi; 170,000 ft). Temperature increases with height. The stratosphere contains the ozone layer, the part of the Earth's atmosphere which contains relatively high concentrations of ozone. "Relatively high" means a few parts per million—much higher than the concentrations in the lower atmosphere but still small compared to the main components of the atmosphere. It is mainly located in the lower portion of the stratosphere from approximately 15–35 km (9.3–22 mi; 49,000–110,000 ft) above Earth's surface, though the thickness varies seasonally and geographically.

Ozone Layer

Though part of the Stratosphere, the ozone layer is considered as a layer of the Earth's atmosphere in itself because its physical and chemical composition is far different from the Stratosphere. Ozone (O_3) in the Earth's stratosphere is created by ultraviolet light striking oxygen molecules containing two oxygen atoms (O_2), splitting them into individual oxygen atoms (atomic oxygen); the atomic oxygen then combines with unbroken O_2 to create O_3. O_3 is unstable (although, in the stratosphere, long-lived) and when ultraviolet light hits ozone it splits into a molecule of O_2 and an atom of atomic oxygen, a continuing process called the ozone-oxygen cycle. This occurs in the ozone layer, the region from about 10 to 50 km (33,000 to 160,000 ft) above Earth's surface. About 90% of the ozone in our atmosphere is contained in the stratosphere. Ozone concentrations are greatest between about 20 and 40 km (66,000 and 130,000 ft), where they range from about 2 to 8 parts per million.

Tropopause

The boundary between the stratosphere and troposphere.

Troposphere

From the Greek word "ôñÝðù" meaning to turn or change. The troposphere is the lowest layer of the atmosphere; it begins at the surface and extends to between 7 km (23,000 ft) at the poles and 17 km (56,000 ft) at the equator, with some variation due to weather factors. The troposphere has a great deal of vertical mixing because of solar heating at the area. This heating makes air masses less dense so they rise. When an air mass rises, the pressure upon it decreases so it expands, doing work against the opposing pressure of the surrounding air. To do work is to expend energy, so the temperature of the air mass decreases. As the temperature decreases, water vapor in the air mass may condense or solidify, releasing latent heat that further uplifts the air mass. This process determines the maximum rate of decline of temperature with height, called the adiabatic lapse rate. The troposphere contains roughly 80% of the total mass of the atmosphere. Fifty per cent of the total mass of the atmosphere is located in the lower 5.6 km (18,000 ft) of the troposphere.

The average temperature of the atmosphere at the surface of Earth is 14°C (57°F; 287 K) or 15°C (59°F; 288 K), depending on the reference.

Pressure and Thickness

The average atmospheric pressure, at sea level, is about 1 atmosphere (atm) = 101.3 kPa (kilopascals) = 14.7 psi (pounds per square inch) = 760 torr = 29.9 inches of mercury (symbol Hg). Total atmospheric mass is 5.1480×10^{18} kg (1.135×10^{19} lb).

Atmospheric pressure is a direct result of the total weight of the air above the point at which the pressure is measured. Air pressure varies with location and time, because the amount (and weight) of air above the earth varies with location and time. However, the *average* mass of the air above a square meter of the Earth's surface can be calculated from the total amount of air and the surface area of the Earth. The total air mass is 5148.0 teratonnes and area is 51007.2 megahectares. Thus 5148.0/510.072 = 10.093 tonnes (9.934 LT; 11.126 ST) per square metre or 14.356 pounds per square inch (98.98 kPa). This is about 2.5% below the officially standardized unit atmosphere (1 atm) of 101.325 kPa or 14.696 psi, and corresponds to the mean pressure not at sea level, but at the mean base of the atmosphere as contoured by the Earth's terrain.

Were atmospheric density to remain constant with height the atmosphere would terminate abruptly at 7.81 km (25,600 ft). Instead, density decreases with height, dropping by 50% at an altitude of about 5.6 km (18,000 ft). For comparison the highest mountain, Mount Everest, is higher, at 8.8 km (29,000 ft), so air is less than half as dense at the summit than at sea level. This is why it is so difficult to climb without supplemental oxygen.

This pressure drop is approximately exponential, so that pressure decreases by approximately half every 5.6 km (18,000 ft) and by 63.2% (1 " 1 / *e* = 1 " 0.368 = 0.632) every 7.64 km (25,100 ft), the average scale height of Earth's atmosphere below 70 km (43 mi; 230,000 ft). However, because of changes in temperature, average molecular weight, and gravity throughout the atmospheric column, the dependence of atmospheric pressure on altitude is modeled by separate equations for each of the layers listed above. Even in the exosphere, the atmosphere is still present. This can be seen by the effects of atmospheric drag on satellites.

In summary, the equations of pressure by altitude in the above references can be used directly to estimate atmospheric thickness. However, the following published data are given for reference:

- 50% of the atmosphere by mass is below an altitude of 5.6 km (18,000 ft).
- 90% of the atmosphere by mass is below an altitude of 16 km (52,000 ft). The common altitude of commercial airliners is about 10 km (33,000 ft) and Mt. Everest's summit is 8,848 m (29,030 ft) above sea level.
- 99.99997% of the atmosphere by mass is below 100 km (62 mi; 330,000 ft), although in the rarefied region above this there are auroras and other atmospheric effects. The highest X-15 plane flight in 1963 reached an altitude of 354,300 ft (108.0 km).

Composition

Filtered air includes trace amounts of many of the chemical elements. Substantial amounts of argon, nitrogen, and oxygen are present as elementary gases. Note the major greenhouse gases: water vapour, carbon dioxide, methane, nitrous oxide, and ozone. Many additional elements from natural sources may be present in tiny amounts in an unfiltered air sample, including contributions from dust, pollen and spores, sea spray, vulcanism, and meteoroids. Various industrial pollutants are also now present in the air, such as chlorine (elementary or in compounds), fluorine (in compounds), elementary mercury, and sulfur (in compounds such as sulfur dioxide).

Composition of dry atmosphere, by volume

Gas	*Volume*
Nitrogen (N_2)	780,840 ppmv (78.084%)
Oxygen (O_2)	209,460 ppmv (20.946%)
Argon (Ar)	9,340 ppmv (0.9340%)
Carbon dioxide (CO_2)	383 ppmv (0.0383%)
Neon (Ne)	18.18 ppmv (0.001818%)
Helium (He)	5.24 ppmv (0.000524%)
Methane (CH_4)	1.745 ppmv (0.0001745%)
Krypton (Kr)	1.14 ppmv (0.000114%)
Hydrogen (H_2)	0.55 ppmv (0.000055%)
Nitrous oxide (N_2O)	0.3 ppmv (0.00003%)
Xenon (Xe)	0.09 ppmv ($9x10^{-6}$%)
Ozone (O_3)	0.0 to 0.07 ppmv (0%-$7x10^{-6}$%)
Nitrogen dioxide (NO_2)	0.02 ppmv ($2x10^{-6}$%)
Iodine (I)	0.01 ppmv ($1x10^{-6}$%)
Carbon monoxide (CO)	0.1 ppmv
Ammonia (NH_3)	trace

Not included in above dry atmosphere:

Water vapour (H_2O)	~0.40% over full atmosphere, typically 1%-4% at surface

PPMV

The *parts per million by volume* figures above are by volume-fraction (V%), which for ideal gases is equal to mole-fraction (that is, the fraction of total molecules). Although the atmosphere is not an ideal gas, nonetheless the atmosphere behaves enough like an ideal gas that the volume-fraction is the same as the mole-fraction for the precision given.

By contrast, *mass-fraction* abundances of gases will differ from the volume values. The mean molar mass of air is 28.97 g/mol, while the molar mass of helium is 4.00, and krypton is 83.80. Thus helium is 5.2 ppm by *volume-fraction*, but 0.72 ppm by *mass-fraction* ([4/29] × 5.2 = 0.72), and krypton is 1.1 ppm by *volume-fraction*, but 3.2 ppm by *mass-fraction* ([84/29] × 1.1 = 3.2).

Heterosphere

Below the *turbopause*, at an altitude of about 100 km (62 mi; 330,000 ft) (not far from the mesopause), the Earth's atmosphere has a more-or-less uniform composition (apart from water vapor) as described above; this constitutes the *homosphere*. However, above the turbopause, the Earth's atmosphere begins to have a composition which varies with altitude. This is because, in the absence of mixing, the density of a gas falls off exponentially with increasing altitude but at a rate which depends on the molar mass. Thus higher mass constituents, such as oxygen and nitrogen, fall off more quickly than lighter constituents such as helium and hydrogen. Thus there is a layer, called the *heterosphere*, in which the Earth's atmosphere has varying composition.

Density and Mass

The density of air at sea level is about 1.2 kg/m^3 (1.2 g/L). Natural variations of the barometric pressure occur at any one altitude as a consequence of weather. This variation is relatively small for inhabited altitudes but much more pronounced in the outer atmosphere and space because of variable solar radiation.

The atmospheric density decreases as the altitude increases. This variation can be approximately modeled using the barometric formula. More sophisticated models are used by meteorologists and space agencies to predict weather and orbital decay of satellites.

The average mass of the atmosphere is about 5 quadrillion metric tons or 1/1,200,000 the mass of Earth. According to the National Center for Atmospheric Research, "The total mean mass of the atmosphere is 5.1480×10^{18} kg with an annual range due to water vapor of 1.2 or 1.5×10^{15} kg depending on whether surface pressure or water vapor data are used; somewhat smaller than the previous estimate. The mean mass of water vapor is estimated as 1.27×10^{16} kg and the dry air mass as $5.1352 \pm 0.0003 \times 10^{18}$ kg."

Optical Properties

Solar radiation (or sunlight) is the energy the Earth receives from the Sun. The Earth also emits radiation back into space, but at longer wavelengths that we cannot see. Depending on its condition, the atmosphere can block radiation from coming in or going out. Important examples of this are clouds and the greenhouse effect.

Scattering

When light passes through our atmosphere, photons interact with it through *scattering*. If the light does not interact with the atmosphere, it is called *direct radiation* and is what you see if you were to look directly at the sun. *Indirect radiation* is light that has been scattered in the atmosphere. For example, on an overcast day when you can't see your shadow there is no direct radiation reaching you, it has all been scattered. As another example, due to a phenomenon called Rayleigh scattering, shorter (blue) wavelengths scatter more easily than longer (red) wavelengths. This is why the sky looks blue, you are seeing scattered blue light. This is also why sunsets are red. Because the sun is close to the horizon, the sun rays pass through more atmosphere than normal to reach your eye. Much of the blue light has been scattered out, leaving the red light in a sunset.

Absorption

Absorption is another important property of the atmosphere. Different molecules absorb different wavelengths of radiation. For example, O_2 and O_3 absorb almost all wavelengths shorter than 300 nanometers. Water (H_2O) absorbs many wavelengths above 700 nm, but this depends on the amount of water vapour in the atmosphere. When a molecule absorbs a photon, it increases the energy of the molecule. We can think of this as heating the atmosphere, but the atmosphere also cools by emitting radiation, as discussed below.

When you combine the absorption spectra of the gasses in the atmosphere, you are left with "windows" of low opacity, allowing the transmission of only certain bands of light. The optical window runs from around 300 nm (ultraviolet-C) up into the range humans can see, the visible spectrum (commonly called light), at roughly 400–700 nm and continues to the infrared to around 1100 nm. There are also infrared and radio windows that transmit some infrared and radio waves at longer wavelengths. For example, the radio window runs from about one centimeter to about eleven-meter waves.

Emission

Emission is the opposite of absorption, it is when an object emits radiation. Objects tend to emit amounts and wavelengths of radiation depending on their "black body" emission curves, therefore hotter objects tend to emit more radiation, with shorter wavelengths. Colder objects emit less radiation, with longer wavelengths. For example, the sun is approximately 6,000 K (5,730°C; 10,340°F), its radiation peaks near 500 nm, and is visible to the human eye. The Earth is approximately 290 K (17°C; 62°F), so its radiation peaks near 10,000 nm, and is much too long to be visible to humans.

Because of its temperature, the atmosphere emits infrared radiation. For example, on clear nights the Earth's surface cools down faster than on cloudy nights. This is because clouds (H_2O) are strong absorbers and emitters of infrared radiation. This is also why it becomes colder at night at higher elevations. The atmosphere acts as a "blanket" to limit the amount of radiation the Earth loses into space.

The *greenhouse effect* is directly related to this absorption and emission (or "blanket") effect. Some chemicals in the atmosphere absorb and emit infrared radiation, but do not interact with sunlight in the visible spectrum. Common examples of these chemicals are CO_2 and H_2O. If there are too much of these *greenhouse gasses,* sunlight heats the Earth's surface, but the gasses block the infrared radiation from exiting back to space. This imbalance causes the Earth to warm, and thus climate change.

Refractive Index

The refractive index of air is close to, but just greater than 1. Systematic variations in refractive index can lead to the bending of light rays over long optical paths. One example is that under some circumstances ships can see other vessels just over the horizon because light is refracted in the same direction as the curvature of the Earth's surface.

Circulation

Atmospheric circulation is the large-scale movement of air, and the means (with ocean circulation) by which heat is distributed on the surface of the Earth.

The large-scale structure of the atmospheric circulation varies from year to year, but the basic structure remains fairly constant. However, individual weather systems - midlatitude depressions, or tropical convective cells - occur "randomly". It is accepted that weather cannot be predicted beyond a fairly short limit; perhaps a month in theory, or about ten days in practice (see Chaos theory and Butterfly effect). Nonetheless, the average of these systems (the climate) is stable over longer periods of time.

Evolution of Earth's Atmosphere

Earliest Atmosphere

The outgassings of the Earth were stripped away by solar wind early in the history of the planet until a steady state was established, the first atmosphere. Based on today's volcanic evidence, this atmosphere would have contained 80% water vapor, 10% carbon dioxide, 5 to 7% hydrogen-sulfur, and smaller amounts of nitrogen, carbon monoxide, hydrogen, methane and inert gases.

A major rainfall led to the buildup of a vast ocean, enriching the other agents, first carbon dioxide and later nitrogen and inert gases. A major part of carbon dioxide exhalations were soon dissolved in water and built up carbonatic sediments.

Second Atmosphere

Water related sediments have been found dating from as early as 3.8 billion years ago . About 3.4 billion years ago, nitrogen was the major part of the then stable "second atmosphere." An influence of life has to be taken into account rather soon in the history of the atmosphere, since hints of early life forms are to be found as early as 3.5 billion years ago. The fact that this is not perfectly in line with the — compared to today 30% lower — solar radiance of the early sun has been described as the "Faint young Sun paradox".

The geological record however shows a continually relatively warm surface during the complete early temperature record of the Earth with the exception of one cold glacial phase about 2.4 billion years ago. Sometime during the late Archaean era an oxygen-containing atmosphere began to develop, apparently from photosynthesizing algae which have been found as stromatolite fossils from 2.7 billion years ago. The early basic carbon isotopy (isotope ratio proportions) is very much in line with what is found today, suggesting that the fundamental features of the carbon cycle were established as early as 4 billion years ago.

Third Atmosphere

The accretion of continents about 3.5 billion years ago added plate tectonics, constantly rearranging the continents and also shaping long-term climate evolution by allowing the transfer of carbon dioxide to large land-based carbonate storages. Free oxygen did not exist until about 1.7 billion years ago and this can be seen with the development of the red beds and the end of the banded iron formations. This signifies a shift from a reducing atmosphere to an oxidizing atmosphere. O_2 showed major ups and downs until reaching a steady state of more than 15%. The following time span was the Phanerozoic era, during which oxygen-breathing metazoan life forms began to appear.

Currently, anthropogenic greenhouse gases are increasing in the atmosphere. According to the Intergovernmental Panel on Climate Change, this increase is a causative factor in global warming.

Air Pollution

Air pollution is the human introduction of chemicals, particulate matter, or biological materials that cause harm or discomfort to organisms, into the atmosphere. Stratospheric ozone depletion is believed to be caused by air pollution (chiefly from chlorofluorocarbons).

Worldwide, air pollution is responsible for large numbers of deaths and respiratory disease. Enforced air quality standards, like the Clean Air Act in the United States, have reduced the presence of some pollutants. While major stationary sources are often identified with air pollution, the greatest source of emissions is actually mobile sources, principally the automobile.

Ashen Light

Ashen light was first sighted by the astronomer Giovanni Battista Riccioli on January 9, 1643, it is a subtle glow that is seen from the night side of the planet Venus. This is comparable to an observer on the Moon seeing China in daylight while it is dark in North America. This Ashen Light is said to be very similar to Earthshine on our moon, but not as distinguished in brightness. It has been frequently sighted by various researchers including Sir William Herschel, Patrick Moore, Dale P. Cruikshank and William K. Hartmann.

Before the development of more powerful telescopes, early astronomer Franz von Gruithuisen believed that Ashen light was from the fires from celebration of a new Venusian emperor, and later believed that it was the inhabitants burning vegetation to make room for farmland.

Detection Attempts

Several space missions have been dedicated to record sightings of the light in hopes of solidifying what Riccioli and others have claimed. The Pioneer and the Russian Venera 11 and 12 sent to Venus did not enjoy success in this area; it is believed that the Ashen Light can only be seen through human eyes. There are a few hypotheses being considered to try to explain this astronomical phenomenon.

An attempt to view the Ashen light was made in Hawaii using the Keck 1 telescope. Researchers claimed to see a faint green glow on the night side of Venus. They suspected that is was carbon dioxide which is known be of a high concentration in that atmosphere. When the molecules are split by the ultraviolet light from the Sun, they become carbon monoxide and oxygen, which emits a green light. However, this light emitted is very faint, and researchers doubt that it is the explanation surrounding Ashen Light.

Amateur astronomers attempting to view the Ashen Light can try it by using an occulting bar, an opaque mask for one's eyes. This lens blocks the sunlit portion of Venus which greatly cuts down extraneous light that is scattered in the eye, improving the chances of witnessing the faint Ashen glow. However, Venus light scattered by the Earth's atmosphere and in the telescope's lens still works against the observer, and it is very hard to get the timing just right. Nevertheless, there have been calculated times in which the light scattered by Earth is covered partially by our Moon. For example, on July 17, 2001, the progression of a crescent Moon temporarily hid the illuminated portion of Venus. Unfortunately the location needed to catch this almost perfect occultation was somewhere in the middle of the Atlantic Ocean, and it was only visible in this manner for 10 to 20 seconds.

Other Hypotheses

Another hypothesis for the cause of the Ashen light is lightning. If there are several strikes in a period of time, the sequence may give off an overall glow in the skies of Venus, if this happens on the night side. However, in an article published in the January issue of Nature, a team of astronomers from the University of Iowa expressed negative criticism and doubt of this idea. After reviewing data taken from the Cassini spacecraft, which flew close to Venus in 1998 and 1999, the team resolved that no high-frequency radio noise was detected. On Earth we hear "static"-like noise on AM radio stations during thunderstorms. If the light was caused by lightning, similar noise would have been expected. In 2007, whistler waves were detected by Venus Express, confirming the occurrence of lightning on Venus. This seems to be the most promising hypothesis, supported by many astronomers, including two researchers whose paper on their findings is examined below.

El Nino-Southern Oscillation

El Nino-Southern Oscillation is a combination of atmospheric and oceanic phenomena in the tropical Pacific Ocean. It is manifested in the atmosphere by changes in the pressure difference between Tahiti and Darwin, Australia, and in the ocean by warming or cooling of surface waters of the tropical Eastern Pacific Ocean. *El Niño* is the name given to the period when water in that region is warmer than normal while *La Niña* is the period when the water there is colder than normal. The oscillation has no well-defined period, but instead occurs every three to eight years. Mechanisms that sustain the El Niño - La Niña cycle remain a matter of research.

The El Niño-Southern Oscillation is often is abbreviated in scientific jargon as ENSO and in popular usage is commonly called simply El Niño. The name is Spanish for "the little boy" and refers to the Christ child, because periodic warming in the Pacific near South America is usually noticed around Christmas.

El Niño is associated with floods, droughts and is linked to other weather disturbances in many locations around the world. El Niño's effects in the Atlantic Ocean lag behind those in the Pacific by 12 to 18 months. Developing countries dependent upon agricultural and fishing are especially affected. El Niño's effects on weather vary with each event. Recent research suggests that treating ocean warming in the eastern tropical Pacific separately from that in the central tropical Pacific may help explain some of these variations.

Causes and Characteristics of El Nino

Although its causes are still being investigated, El Niño events begin when trade winds, part of the Walker circulation, falter for many months. A series of Kelvin waves—relatively warm subsurface waves of water a few centimeters high and hundreds of kilometres wide—cross the Pacific along the equator and create a pool of warm water near South America, where ocean temperatures are normally cold due to upwelling. The Pacific Ocean is a heat reservoir that drives global wind patterns, and the resulting change in its temperature alters weather on a global scale. Rainfall shifts from the western Pacific toward the Americas, while Indonesia and India become drier.

Jacob Bjerknes in 1969 helped toward an understanding of ENSO, by suggesting that an anomalously warm spot in the eastern Pacific can weaken the east-west temperature difference, disrupting trade winds that push warm water to the west. The result is increasingly warm water toward the east. Several mechanisms have been proposed through which warmth builds up in equatorial Pacific surface waters, and is then dispersed to lower depths by an El Niño event. The resulting cooler area then has to "recharge" warmth for several years before another event can take place.

While not a direct cause of El Niño, the Madden-Julian Oscillation, or MJO, propagates rainfall eastward around the global tropics in a cycle of 30–60 days, and may influence the speed of development and intensity of El Niño and La Niña in several ways. For example, westerly flows between MJO-induced areas of low pressure may cause cyclonic circulations north and south of the equator. When the circulations intensify, the westerly winds within the equatorial Pacific can further increase and shift eastward, playing a role in El Niño development. Madden-Julian activity can also produce eastward-propagating oceanic Kelvin waves, which may in turn be influenced by a developing El Niño, leading to a positive feedback loop.

Definition

Climate scientists define El Nino and La Nina based on a sustained differences in Pacific-Ocean surface temperatures when compared with the average value. The accepted difference is anything greater than 0.5C (or 0.9F) averaged over the east-central tropical Pacific Ocean. When this happens for less than five months, it is classified as El Niño or La Nina conditions; if the anomaly persists for five months or longer, it is called an El Nino or La Nina "episode." Typically, this happens at irregular intervals of 2–7 years and lasts nine months to two years.

The first signs of an El Nino are:

1. Rise in surface pressure over the Indian Ocean, Indonesia, and Australia
2. Fall in air pressure over Tahiti and the rest of the central and eastern Pacific Ocean
3. Trade winds in the south Pacific weaken or head east
4. Warm air rises near Peru, causing rain in the northern Peruvian deserts
5. Warm water spreads from the west Pacific and the Indian Ocean to the east Pacific. It takes the rain with it, causing extensive drought in the western Pacific and rainfall in the normally dry eastern Pacific.

El Nino's warm current of nutrient-poor tropical water, heated by its eastward passage in the Equatorial Current, replaces the cold, nutrient-rich surface water of the Humboldt

Current. When El Niño conditions last for many months, extensive ocean warming occurs and its economic impact to local fishing for an international market can be serious.

Southern Oscillation

The Southern Oscillation is the atmospheric counterpart of El Niño. It is an oscillation in air pressure between the tropical eastern and the western Pacific Ocean waters. The strength of the Southern Oscillation is measured by the *Southern Oscillation Index* (SOI). The SOI is computed from fluctuations in the normalized surface air pressure difference between Tahiti and Darwin, Australia. El Niño episodes are associated with negative values of the SOI, meaning that the pressure at Tahiti is relatively low compared to Darwin.

Low atmospheric pressure tends to occur over warm water and cold pressure occurs over cold water, in part because deep convection over the warm water acts to transport air. El Nino episodes are defined as sustained warming of the central and eastern tropical Pacific Ocean. This results in a decrease in the strength of the Pacific trade winds, and a reduction in rainfall over eastern and northern Australia. Conversely, La Nina episodes are associated with *positive values* of the SOI and are accompanied by stronger Pacific trade winds and warmer sea temperatures to the north of Australia. Waters in the central and eastern tropical Pacific Ocean become cooler during this time.

Walker Circulation

During non-El Niño conditions, the Walker circulation is seen at the surface as easterly trade winds which move water and air warmed by the sun towards the west. This also creates ocean upwelling off the coasts of Peru and Ecuador and brings nutrient-rich cold water to the surface, increasing fishing stocks. The western side of the equatorial Pacific is characterized by warm, wet low pressure weather as the collected moisture is dumped in the form of typhoons and thunderstorms. The ocean is some 60 centimetres (24 in) higher in the western Pacific as the result of this motion.

In the Pacific, La Niña is characterized by unusually cold ocean temperatures in the eastern equatorial Pacific, compared to El Niño, which is characterized by unusually warm ocean temperatures in the same area. The La Niña condition and El Niño condition alternate over a several year cycle.

Tropical Cyclones

Most tropical cyclones form on the side of the subtropical ridge closer to the equator, then move poleward past the ridge axis before recurving into the main belt of the Westerlies. When the subtropical ridge position shifts due to El Nino, so will the preferred tropical cyclone tracks. Areas west of Japan and Korea tend to experience much fewer September-November tropical cyclone impacts during El Niño and neutral years. During El Niño years, the break in the subtropical ridge tends to lie near 130°E which would favor the Japanese archipelago. During El Niño years, Guam's chance of a tropical cyclone impact is one-third of the long term average. The tropical Atlantic ocean experiences depressed activity due to increased vertical wind shear across the region during El Nino years.

Elsewhere

In Africa, East Africa, including Kenya, Tanzania and the White Nile basin experiences, in the long rains from March to May, wetter than normal conditions. There also are drier than normal conditions from December to February in south-central Africa, mainly in Zambia, Zimbabwe, Mozambique and Botswana. Direct effects of El Niño resulting in drier conditions occur in parts of Southeast Asia and Northern Australia, increasing bush fires and worsening haze and decreasing air quality dramatically. Drier than normal conditions are also generally observed in Queensland, inland Victoria, inland New South Wales and eastern Tasmania from June to August. West of the Antarctic Peninsula, the Ross, Bellingshausen, and Amundsen Sea sectors have more sea ice during El Niño. The latter two and the Weddell Sea also become warmer and have higher atmospheric pressure.

El Nino's effects on Europe are not entirely clear. There is some evidence that an El Nino may cause a wetter, cloudier winter in Northern Europe and a milder, drier winter in the Mediterranean Sea region. However, the El Nino winter of 2006/2007 was unusually mild in the UK and Western Europe, and the Alps recorded very little snow coverage that season.

La Nina

La Nina is the name for the cold phase of ENSO, during which the cold pool in the eastern Pacific intensifies and the trade winds strengthen. The name La Nina originates from Spanish, meaning "the little girl", analogous to El Nino meaning "the little boy". It has also in the past been called *anti-El Nino,* and El Viejo.

North America

La Niña causes mostly the opposite effects of El Nino. Atlantic tropical cyclone activity is generally enhanced during La Nina. La Nina causes increased rainfall across the United States' Midwest. Other potential impacts include above average precipitation in the Northern Rockies, Northern California, and in southern and eastern regions of the Pacific Northwest. Below-average precipitation is expected across the southern tier, particularly in the southwestern and southeastern states."

Asia

During La Niña years, the formation of tropical cyclones, along with the subtropical ridge position, shifts westward across the western Pacific ocean, which increases the landfall threat to China. In March 2008, La Nina caused a drop in sea surface temperatures over Southeast Asia by an amount of 2°C. It also caused heavy rains over Malaysia, Philippines and Indonesia.

Recent Occurrences

There was a strong La Niña episode during 1988-1989. La Niña also formed in 1995, from 1998-2000, and a minor one from 2000-2001. The most recent occurrence of El Nino started in September 2006 and lasted until early 2007. From June 2007 on, data indicated a moderate La Nina event, which strengthened in early 2008 and weakened by early 2009. According to NOAA, El Niño conditions have been in place in the equatorial Pacific Ocean since June 2009. Therefore the 2009/2010 season is expected to be an El Niño year.

Atlantic Counterpart

An effect similar to El Nino sometimes takes place in the Atlantic Ocean, where water along equatorial Africa's Gulf of Guinea becomes warmer and eastern Brazil becomes cooler and drier. This is related to El Nino's effect on the Walker circulation over South America, which causes the easterly trade winds in the western Atlantic Ocean region to strengthen. Study of climate records has found that about half of the summers after an El Nino have unusual warming in the Western Hemisphere Warm Pool (WHWP). This affects weather in the area and seems to be related to the North Atlantic Oscillation. Cases of El Niño-type events in both oceans simultaneously have been linked to severe famines related to the extended failure of monsoon rains.

ENSO and Global Warming

A few years ago, attribution of recent changes (if any) in ENSO or predictions of future changes were very weak. More recent results tend to suggest that the projected tropical warming may follow a somewhat El Nino-like spatial pattern, without necessarily altering the variability about this pattern, while the ENSO cycle may be minimally shortened.

El Nino "Modoki"

One recent study suggests that El Niños have been changing. This study says there are two similar forms of warming in the Pacific Ocean. One is the eastern Pacific Warming, which is the same as a typical El Niño. The central Pacific Warming has temperature anomalies near the dateline rather than near South America. There is some reason to believe that this warming in the central Pacific reduces El Niño's usual suppression of Atlantic hurricane activity. Warming in the central Pacific is typical of La Ninas. This new type of El Nino has been dubbed "El Nino Modoki". The change from El Nino to El Nino Modoki may be due to a weakening of the southeasterly trade winds. It is not known whether this is due to natural variablity, the effects of climate change, or some other cause. Research is ongoing to determine whether a similar thing is happening to La Nina.

Cultural History and Pre-historic Information

ENSO conditions have occurred at two- to seven year intervals for at least the past 300 years, but most of them have been weak. There is also evidence for strong El Nino events during the early Holocene epoch 10,000 years ago.

El Nino affected pre-Columbian Incas and may have led to the demise of the Moche and other pre-Columbian Peruvian cultures. A recent study suggests that a strong El-Nino effect between 1789-93 caused poor crop yields in Europe, which in turn helped touch off the French Revolution.

An early recorded mention of the term "El Nino" to refer to climate occurs in 1892, when Captain Camilo Carrillo told the Geographical society congress in Lima that Peruvian sailors named the warm northerly current "El Nino" because it was most noticeable around Christmas. The phenomenon had long been of interest because of its effects on the guano industry and other enterprises that depend on biological productivity of the sea.

Charles Todd, in 1893, suggested that droughts in India and Australia tended to occur at the same time; Norman Lockyer noted the same in 1904. An El Nino connection with flooding was reported in 1895 by Pezet and Eguiguren. In 1924 Gilbert Walker (for whom the Walker circulation is named) coined the term "Southern Oscillation".

The major 1982-83 El Niño lead to an upsurge of interest from the scientific community. The period from 1990-1994 was unusual in that El Ninos have rarely occurred in such rapid succession. An especially intense El Nino event in 1998 caused an estimated 16% of the world's reef systems to die. The event temporarily warmed air temperature by 1.5°C, compared to the usual increase of 0.25°C associated with El Niño events. Since then, mass coral bleaching has become common worldwide, with all regions having suffered 'severe bleaching'. Global warming may according to some evidence increase the intensity and/or frequency of El Nino episodes.

8
Environmental Pollution

Throughout history from Ancient Greece to Andalusia, Ancient China, Central Europe during the Renaissance until today, philosophers ranging from Aristotle, Al-Farabi, Al-Ghazali, Averroes, Buddha, Confucius, Dante, Hegel, Avicenna, Lao Tse, Maimonedes, Montesquieu, Nussbaum, Plato, Socrates and Sun Tzu wrote about the pollution of the body as well as the mind and soul.

Prehistory

Humankind has had some effect upon the environment since the Paleolithic era during which the ability to generate fire was acquired. In the Iron Age, the use of tooling led to the practice of metal grinding on a small scale and resulted in minor accumulations of discarded material probably easily dispersed without too much impact. Human wastes would have polluted rivers or water sources to some degree. However, these effects could be expected predominantly to be dwarfed by the natural world.

Ancient Cultures

The first advanced civilizations of Mesopotamia, Egypt, India, China, Persia, Greece and Rome increased the use of water for their manufacture of goods, increasingly forged metal and created fires of wood and peat for more elaborate purposes (for example, bathing, heating). The forging of metals appears to be a key turning point in the creation of significant air pollution levels. Core samples of glaciers in

Greenland indicate increases in air pollution associated with Greek, Roman and Chinese metal production. Still, at this time the scale of higher activity probably did not disrupt ecosystems.

Middle Ages

The European Dark Ages during the early Middle Ages probably saw a reprieve in widespread pollution, in that industrial activity fell, and population levels did not grow rapidly. Toward the end of the Middle Ages populations grew and concentrated more within cities, creating pockets of readily evident contamination. In certain places air pollution levels were recognizable as health issues, and water pollution in population centers was a serious medium for disease transmission from untreated human waste.

Since travel and widespread information were less common, there did not exist a more general context than that of local consequences in which to consider pollution. Air pollution was largely from wood burning which must be properly ventilated. Septic contamination or poisoning of a clean drinking water source was very easily fatal, and contamination was not well understood. Bad septic contamination and pollution contributed greatly to the Bubonic plague.

But gradually increasing populations and the proliferation of basic industrial processes saw the emergence of a civilization that began to have a much greater collective impact on its surroundings. It was to be expected that the beginnings of environmental awareness would occur in the more developed cultures, particularly in the densest urban centers. The first medium warranting official policy measures in the emerging western world would be the most basic: the air we breathe.

The earliest known writings concerned with pollution were Arabic medical treatises written between the 9th and

13th centuries, by physicians such as al-Kindi (Alkindus), Qusta ibn Luqa (Costa ben Luca), Muhammad ibn Zakarîya Râzi (Rhazes), Ibn Al-Jazzar, al-Tamimi, al-Masihi, Ibn Sina (Avicenna), Ali ibn Ridwan, Ibn Jumay, Isaac Israeli ben Solomon, Abd-el-Latif, Ibn al-Quff, and Ibn al-Nafis. Their works covered a number of subjects related to pollution such as air contamination, water contamination, soil contamination, solid waste mishandling, and environmental assessments of certain localities.

King Edward I of England banned the burning of sea-coal by proclamation in London in 1272, after its smoke had become a problem. But the fuel was so common in England that this earliest of names for it was acquired because it could be carted away from some shores by the wheelbarrow. Air pollution would continue to be a problem in England, especially later during the industrial revolution, and extending into the recent past with the Great Smog of 1952. This same city also recorded one of the earlier extreme cases of water quality problems with the Great Stink on the Thames of 1858, which led to construction of the London sewerage system soon afterward.

It was the industrial revolution that gave birth to environmental pollution as we know it today. The emergence of great factories and consumption of immense quantities of coal and other fossil fuels gave rise to unprecedented air pollution and the large volume of industrial chemical discharges added to the growing load of untreated human waste. Chicago and Cincinnati were the first two American cities to enact laws ensuring cleaner air in 1881. Other cities followed around the country until early in the 20th century, when the short lived Office of Air Pollution was created under the Department of the Interior. Extreme smog events were experienced by the cities of Los Angeles and Donora, Pennsylvania in the late 1940s, serving as another public reminder.

Modern Awareness

Pollution became a popular issue after World War-II, when the aftermath of atomic warfare and testing made evident the perils of radioactive fallout. Then a conventional catastrophic event The Great Smog of 1952 in London killed at least 8000 people. This massive event prompted some of the first major modern environmental legislation, The Clean Air Act of 1956.

Pollution began to draw major public attention in the United States between the mid-1950s and early 1970s, when Congress passed the Noise Control Act, the Clean Air Act, the Clean Water Act and the National Environmental Policy Act.

Bad bouts of local pollution helped increase consciousness. PCB dumping in the Hudson River resulted in a ban by the EPA on consumption of its fish in 1974. Long-term dioxin contamination at Love Canal starting in 1947 became a national news story in 1978 and led to the Superfund legislation of 1980. Legal proceedings in the 1990s helped bring to light Chromium-6 releases in California—the champions of whose victims became famous. The pollution of industrial land gave rise to the name brownfield, a term now common in city planning. DDT was banned in most of the developed world after the publication of Rachel Carson's *Silent Spring*.

The development of nuclear science introduced radioactive contamination, which can remain lethally radioactive for hundreds of thousands of years. Lake Karachay, named by the Worldwatch Institute as the "most polluted spot" on earth, served as a disposal site for the Soviet Union throughout the 1950s and 1960s. Second place may go to the area of Chelyabinsk U.S.S.R. as the "Most polluted place on the planet".

Nuclear weapons continued to be tested in the Cold War, sometimes near inhabited areas, especially in the earlier stages of their development. The toll on the worst-affected populations and the growth since then in understanding about the critical threat to human health posed by radioactivity has also been a prohibitive complication associated with nuclear power. Though extreme care is practiced in that industry, the potential for disaster suggested by incidents such as those at Three Mile Island and Chernobyl pose a lingering specter of public mistrust. One legacy of nuclear testing before most forms were banned has been significantly raised levels of background radiation.

International catastrophes such as the wreck of the Amoco Cadiz oil tanker off the coast of Brittany in 1978 and the Bhopal disaster in 1984 have demonstrated the universality of such events and the scale on which efforts to address them needed to engage. The borderless nature of atmosphere and oceans inevitably resulted in the implication of pollution on a planetary level with the issue of global warming. Most recently the term persistent organic pollutant (POP) has come to describe a group of chemicals such as PBDEs and PFCs among others. Though their effects remain somewhat less well understood owing to a lack of experimental data, they have been detected in various ecological habitats far removed from industrial activity such as the Arctic, demonstrating diffusion and bioaccumulation after only a relatively brief period of widespread use.

Growing evidence of local and global pollution and an increasingly informed public over time have given rise to environmentalism and the environmental movement, which generally seek to limit human impact on the environment.

Forms of Pollution

The major forms of pollution are listed below along with the particular pollutants relevant to each of them:

- Air pollution, the release of chemicals and particulates into the atmosphere. Common gaseous air pollutants include carbon monoxide, sulfur dioxide, chlorofluorocarbons (CFCs) and nitrogen oxides produced by industry and motor vehicles. Photochemical ozone and smog are created as nitrogen oxides and hydrocarbons react to sunlight. Particulate matter, or fine dust is characterized by their micrometre size PM_{10} to $PM_{2.5}$.

- Water pollution, by the release of waste products and contaminants into surface runoff into river drainage systems, leaching into groundwater, liquid spills, wastewater discharges, eutrophication and littering.
- Soil contamination occurs when chemicals are released by spill or underground leakage. Among the most significant soil contaminants are hydrocarbons, heavy metals, MTBE, herbicides, pesticides and chlorinated hydrocarbons.
- Littering
- Radioactive contamination, resulting from 20th century activities in atomic physics, such as nuclear power generation and nuclear weapons research, manufacture and deployment. (See alpha emitters and actinides in the environment.)
- Noise pollution, which encompasses roadway noise, aircraft noise, industrial noise as well as high-intensity sonar.
- Light pollution, includes light trespass, over-illumination and astronomical interference.
- Visual pollution, which can refer to the presence of overhead power lines, motorway billboards, scarred landforms (as from strip mining), open storage of trash or municipal solid waste.
- Thermal pollution, is a temperature change in natural water bodies caused by human influence, such as use of water as coolant in a power plant.

AIR POLLUTION

Air pollution is the introduction of chemicals, particulate matter, or biological materials that cause harm or discomfort to humans or other living organisms, or damages the natural environment, into the atmosphere.

The atmosphere is a complex, dynamic natural gaseous system that is essential to support life on planet Earth. Stratospheric ozone depletion due to air pollution has long been recognized as a threat to human health as well as to the Earth's ecosystems.

Pollutants

An air pollutant is known as a substance in the air that can cause harm to humans and the environment. Pollutants can be in the form of solid particles, liquid droplets, or gases. In addition, they may be natural or man-made.

Pollutants can be classified as either primary or secondary. Usually, primary pollutants are substances directly emitted from a process, such as ash from a volcanic eruption, the carbon monoxide gas from a motor vehicle exhaust or sulfur dioxide released from factories.

Secondary pollutants are not emitted directly. Rather, they form in the air when primary pollutants react or interact. An important example of a secondary pollutant is ground level ozone - one of the many secondary pollutants that make up photochemical smog.

Note that some pollutants may be both primary and secondary: that is, they are both emitted directly and formed from other primary pollutants. About 4 per cent of deaths in the United States can be attributed to air pollution, according to the Environmental Science Engineering Programme at the Harvard School of Public Health.

Major primary pollutants produced by human activity include:

- Sulfur oxides (SO_x) — Especially sulfur dioxide, a chemical compound with the formula SO_2. SO_2 is produced by volcanoes and in various industrial processes. Since coal and petroleum often contain sulfur compounds, their combustion generates sulfur dioxide. Further oxidation of SO_2, usually in the presence of a catalyst such as NO_2, forms H_2SO_4, and thus acid rain. This is one of the causes for concern over the environmental impact of the use of these fuels as power sources.
- Nitrogen oxides (NO_x) — Especially nitrogen dioxide are emitted from high temperature combustion. Can be seen as the brown haze dome above or plume downwind of cities.Nitrogen dioxide is the chemical compound with the formula NO_2. It is one of the several nitrogen oxides. This reddish-brown toxic gas has a characteristic sharp, biting odor. NO_2 is one of the most prominent air pollutants.
- Carbon monoxide — It is a colourless, odourless, non-irritating but very poisonous gas. It is a product by incomplete combustion of fuel such as natural gas, coal or wood. Vehicular exhaust is a major source of carbon monoxide.
- Carbon dioxide (CO_2) — A greenhouse gas emitted from combustion but is also a gas vital to living organisms. It is a natural gas in the atmosphere.
- Volatile organic compounds — VOCs are an important outdoor air pollutant. In this field they are often divided into the separate categories of methane (CH_4) and non-methane (NMVOCs). Methane is an extremely efficient greenhouse gas which contributes to enhanced global warming. Other hydrocarbon VOCs are also

significant greenhouse gases via their role in creating ozone and in prolonging the life of methane in the atmosphere, although the effect varies depending on local air quality. Within the NMVOCs, the aromatic compounds benzene, toluene and xylene are suspected carcinogens and may lead to leukemia through prolonged exposure. 1,3-butadiene is another dangerous compound which is often associated with industrial uses.

- Particulate matter – Particulates, alternatively referred to as particulate matter (PM) or fine particles, are tiny particles of solid or liquid suspended in a gas. In contrast, aerosol refers to particles and the gas together. Sources of particulate matter can be man made or natural. Some particulates occur naturally, originating from volcanoes, dust storms, forest and grassland fires, living vegetation, and sea spray. Human activities, such as the burning of fossil fuels in vehicles, power plants and various industrial processes also generate significant amounts of aerosols. Averaged over the globe, anthropogenic aerosols—those made by human activities—currently account for about 10 percent of the total amount of aerosols in our atmosphere. Increased levels of fine particles in the air are linked to health hazards such as heart disease, altered lung function and lung cancer.
- Toxic metals, such as lead, cadmium and copper.
- Chlorofluorocarbons (CFCs) – Harmful to the ozone layer emitted from products currently banned from use.
- Ammonia (NH_3) – Emitted from agricultural processes. Ammonia is a compound with the formula NH_3. It is normally encountered as a gas with a characteristic pungent odour. Ammonia contributes significantly to the nutritional needs of terrestrial

organisms by serving as a precursor to foodstuffs and fertilizers. Ammonia, either directly or indirectly, is also a building block for the synthesis of many pharmaceuticals. Although in wide use, ammonia is both caustic and hazardous.

- Odours – Such as from garbage, sewage, and industrial processes
- Radioactive pollutants – produced by nuclear explosions, war explosives, and natural processes such as the radioactive decay of radon.

Secondary pollutants include:

- Particulate matter formed from gaseous primary pollutants and compounds in photochemical smog. Smog is a kind of air pollution; the word "smog" is a portmanteau of smoke and fog. Classic smog results from large amounts of coal burning in an area caused by a mixture of smoke and sulfur dioxide. Modern smog does not usually come from coal but from vehicular and industrial emissions that are acted on in the atmosphere by sunlight to form secondary pollutants that also combine with the primary emissions to form photochemical smog.

- Ground level ozone (O_3) formed from NO_x and VOCs. Ozone (O_3) is a key constituent of the troposphere (it is also an important constituent of certain regions of the stratosphere commonly known as the Ozone layer). Photochemical and chemical reactions involving it drive many of the chemical processes that occur in the atmosphere by day and by night. At abnormally high concentrations brought about by human activities (largely the combustion of fossil fuel), it is a pollutant, and a constituent of smog.

- Peroxyacetyl nitrate (PAN) — Similarly formed from NO_x and VOCs.

Minor air pollutants include:

- A large number of minor hazardous air pollutants. Some of these are regulated in USA under the Clean Air Act and in Europe under the Air Framework Directive.
- A variety of persistent organic pollutants, which can attach to particulate matter.

Persistent Organic Pollutants (POPs) are organic compounds that are resistant to environmental degradation through chemical, biological, and photolytic processes. Because of this, they have been observed to persist in the environment, to be capable of long-range transport, bioaccumulate in human and animal tissue, biomagnify in food chains, and to have potential significant impacts on human health and the environment.

Sources

Sources of air pollution refer to the various locations, activities or factors which are responsible for the releasing of pollutants in the atmosphere. These sources can be classified into two major categories which are:

Anthropogenic sources (human activity) mostly related to burning different kinds of fuel

- "Stationary Sources" include smoke stacks of power plants, manufacturing facilities (factories) and waste incinerators, as well as furnaces and other types of fuel-burning heating devices.
- "Mobile Sources" include motor vehicles, marine vessels, aircraft and the effect of sound etc.

- Chemicals, dust and controlled burn practices in agriculture and forestry management. Controlled or prescribed burning is a technique sometimes used in forest management, farming, prairie restoration or greenhouse gas abatement. Fire is a natural part of both forest and grassland ecology and controlled fire can be a tool for foresters. Controlled burning stimulates the germination of some desirable forest trees, thus renewing the forest.
- Fumes from paint, hair spray, varnish, aerosol sprays and other solvents.
- Waste deposition in landfills, which generate methane.Methane is not toxic; however, it is highly flammable and may form explosive mixtures with air. Methane is also an asphyxiant and may displace oxygen in an enclosed space. Asphyxia or suffocation may result if the oxygen concentration is reduced to below 19.5% by displacement.
- Military, such as nuclear weapons, toxic gases, germ warfare and rocketry.

Natural Sources

- Dust from natural sources, usually large areas of land with little or no vegetation.
- Methane, emitted by the digestion of food by animals, for example, cattle.
- Radon gas from radioactive decay within the Earth's crust. Radon is a colorless, odorless, naturally occurring, radioactive noble gas that is formed from the decay of radium. It is considered to be a health hazard. Radon gas from natural sources can accumulate in buildings, especially in confined areas

such as the basement and it is the second most frequent cause of lung cancer, after cigarette smoking.

- Smoke and carbon monoxide from wildfires.
- Volcanic activity, which produce sulfur, chlorine, and ash particulates.

Emission Factors

Air pollutant emission factors are representative values that attempt to relate the quantity of a pollutant released to the ambient air with an activity associated with the release of that pollutant. These factors are usually expressed as the weight of pollutant divided by a unit weight, volume, distance, or duration of the activity emitting the pollutant (e.g., kilograms of particulate emitted per megagram of coal burned). Such factors facilitate estimation of emissions from various sources of air pollution. In most cases, these factors are simply averages of all available data of acceptable quality, and are generally assumed to be representative of long-term averages.

The United States Environmental Protection Agency has published a compilation of air pollutant emission factors for a multitude of industrial sources. The United Kingdom, Australia, Canada and many other countries have published similar compilations, as well as the European Environment Agency.

Indoor Air Quality (IAQ)

A lack of ventilation indoors concentrates air pollution where people often spend the majority of their time. Radon (Rn) gas, a carcinogen, is exuded from the Earth in certain locations and trapped inside houses. Building materials including carpeting and plywood emit formaldehyde (H_2CO) gas. Paint and solvents give off volatile organic compounds (VOCs) as they dry. Lead paint can degenerate into dust and be inhaled. Intentional air pollution is introduced with the use of air fresheners, incense, and other scented items. Controlled wood fires in stoves and fireplaces can add significant amounts of smoke particulates into the air, inside and out. Indoor pollution fatalities may be caused by using pesticides and other chemical sprays indoors without proper ventilation.

Carbon monoxide (CO) poisoning and fatalities are often caused by faulty vents and chimneys, or by the burning of charcoal indoors. Chronic carbon monoxide poisoning can result even from poorly adjusted pilot lights. Traps are built into all domestic plumbing to keep sewer gas, hydrogen sulfide, out of interiors. Clothing emits tetrachloroethylene, or other dry cleaning fluids, for days after dry cleaning.

Though its use has now been banned in many countries, the extensive use of asbestos in industrial and domestic environments in the past has left a potentially very dangerous material in many localities. Asbestosis is a chronic inflammatory medical condition affecting the tissue of the lungs. It occurs after long-term, heavy exposure to asbestos from asbestos-containing materials in structures. Sufferers have severe dyspnea (shortness of breath) and are at an increased risk regarding several different types of lung cancer. As clear explanations are not always stressed in non-technical literature, care should be taken to distinguish between several forms of relevant diseases. According to the World Health Organization (WHO), these may defined as; asbestosis, *lung cancer*, and *mesothelioma* (generally a very rare form of cancer, when more widespread it is almost always associated with prolonged exposure to asbestos).

Biological sources of air pollution are also found indoors, as gases and airborne particulates. Pets produce dander, people produce dust from minute skin flakes and decomposed hair, dust mites in bedding, carpeting and furniture produce enzymes and micrometre-sized fecal droppings, inhabitants emit methane, mold forms in walls and generates mycotoxins and spores, air conditioning systems can incubate Legionnaires' disease and mold, and houseplants, soil and surrounding gardens can produce pollen, dust, and mold. Indoors, the lack of air circulation allows these airborne pollutants to accumulate more than they would otherwise occur in nature.

Health Effects

The World Health Organization states that 2.4 million people die each year from causes directly attributable to air pollution, with 1.5 million of these deaths attributable to indoor air pollution. "Epidemiological studies suggest that more than 500,000 Americans die each year from cardiopulmonary disease linked to breathing fine particle air pollution. . ." A study by the University of Birmingham has shown a strong correlation between pneumonia related deaths and air pollution from motor vehicles. Worldwide more deaths per year are linked to air pollution than to automobile accidents. Published in 2005 suggests that 310,000 Europeans die from air pollution annually. Direct causes of air pollution related deaths include aggravated asthma, bronchitis, emphysema, lung and heart diseases, and respiratory allergies. The US EPA estimates that a proposed set of changes in diesel engine technology (*Tier* 2) could result in 12,000 fewer *premature mortalities*, 15,000 fewer heart attacks, 6,000 fewer emergency room visits by children with asthma, and 8,900 fewer respiratory-related hospital admissions each year in the United States.

The worst short term civilian pollution crisis in India was the 1984 Bhopal Disaster. Leaked industrial vapours from the Union Carbide factory, belonging to Union Carbide, Inc., U.S.A., killed more than 2,000 people outright and injured anywhere from 150,000 to 600,000 others, some 6,000 of whom would later die from their injuries. The United Kingdom suffered its worst air pollution event when the

December 4 Great Smog of 1952 formed over London. In six days more than 4,000 died, and 8,000 more died within the following months. An accidental leak of anthrax spores from a biological warfare laboratory in the former USSR in 1979 near Sverdlovsk is believed to have been the cause of hundreds of civilian deaths. The worst single incident of air pollution to occur in the United States of America occurred in Donora, Pennsylvania in late October, 1948, when 20 people died and over 7,000 were injured.

The health effects caused by air pollutants may range from subtle biochemical and physiological changes to difficulty in breathing, wheezing, coughing and aggravation of existing respiratory and cardiac conditions. These effects can result in increased medication use, increased doctor or emergency room visits, more hospital admissions and premature death. The human health effects of poor air quality are far reaching, but principally affect the body's respiratory system and the cardiovascular system. Individual reactions to air pollutants depend on the type of pollutant a person is exposed to, the degree of exposure, the individual's health status and genetics.

A new economic study of the health impacts and associated costs of air pollution in the Los Angeles Basin and San Joaquin Valley of Southern California shows that more than 3800 people die prematurely (approximately 14 years earlier than normal) each year because air pollution levels violate federal standards. The number of annual premature deaths is considerably higher than the fatalities related to auto collisions in the same area, which average fewer than 2,000 per year.

Diesel exhaust (DE) is a major contributor to combustion derived particulate matter air pollution. In several human experimental studies, using a well validated exposure chamber setup, DE has been linked to acute vascular dysfunction and increased thrombus formation. This serves as a plausible mechanistic link between the previously described association between particulate matter air pollution and increased cardiovascular morbidity and mortality.

NOISE POLLUTION

Noise pollution (or environmental noise) is displeasing human-, animal- or machine-created sound that disrupts the activity or balance of human or animal life. A common form of noise pollution is from transportation, principally motor vehicles. The word *noise* comes from the Latin word *nausea* meaning seasickness.

The source of most noise worldwide is transportation systems, motor vehicle noise, but also including aircraft noise and rail noise. Poor urban planning may give rise to noise pollution, since side-by-side industrial and residential buildings can result in noise pollution in the residential area.

Other sources are car alarms, emergency service sirens, office equipment, factory machinery, construction work, groundskeeping equipment, barking dogs, appliances, power tools, lighting hum, audio entertainment systems, loudspeakers and noisy people.

Human Health Effects

Noise health effects are both health and behavioural in nature. The unwanted sound is called noise. This unwanted sound can damage physiological and psychological health. Noise pollution can cause annoyance and aggression, hypertension, high stress levels, tinnitus, hearing loss, sleep disturbances, and other harmful effects. Furthermore, stress and hypertension are the leading causes to health problems, whereas tinnitus can lead to forgetfulness, severe depression and at times panic attacks.

Chronic exposure to noise may cause noise-induced hearing loss. Older males exposed to significant occupational noise demonstrate significantly reduced hearing sensitivity than their non-exposed peers, though differences in hearing sensitivity decrease with time and the two groups are indistinguishable by age 79. A comparison of Maaban tribesmen, who were insignificantly exposed to transportation or industrial noise, to a typical U.S. population showed that chronic exposure to moderately high levels of environmental noise contributes to hearing loss.

High noise levels can contribute to cardiovascular effects and exposure to moderately high levels during a single eight hour period causes a statistical rise in blood pressure of five to ten points and an increase in stress and vasoconstriction leading to the increased blood pressure noted above as well as to increased incidence of coronary artery disease.

Noise pollution is also a cause of annoyance. A 2005 study by Spanish researchers found that in urban areas households are willing to pay approximately four Euros per decibel per year for noise reduction.

Environmental Effects

Noise can have a detrimental effect on animals by causing stress, increasing risk of mortality by changing the delicate balance in predator/prey detection and avoidance, and by interfering with their use of sounds in communication especially in relation to reproduction and in navigation. Acoustic overexposure can lead to temporary or permanent loss of hearing.

An impact of noise on animal life is the reduction of usable habitat that noisy areas may cause, which in the case of endangered species may be part of the path to extinction. One of the best known cases of damage caused by noise pollution is the death of certain species of beached whales, brought on by the loud sound of military sonar.

Noise also makes species communicate louder, which is called Lombard vocal response. Scientists and researchers have conducted experiments that show whales' song length is longer when submarine-detectors are on. If creatures don't "speak" loud enough, their voice will be masked by anthropogenic sounds. These unheard voices might be warnings, finding of prey, or preparations of net-bubbling. When one species begins speaking louder, it will mask other species' voice, causing the whole ecosystem to eventually speak louder.

European Robins living in urban environments are more likely to sing at night in places with high levels of noise pollution during the day, suggesting that they sing at night because it is quieter, and their message can propagate through the environment more clearly. Interestingly, the same study showed that daytime noise was a stronger predictor of nocturnal singing than night-time Light pollution, to which the phenomenon is often attributed.

Zebra finches become less faithful to their partners when exposed to traffic noise. This could alter a population's evolutionary trajectory by selecting traits, sapping resources normally devoted to other activities and thus lead to profound genetic and evolutionary consequences.

WATER POLLUTION

Water pollution is the contamination of water bodies such as lakes, rivers, oceans, and groundwater by human activities. All water pollution affects organisms and plants that live in these water bodies and in almost all cases the effect is damaging either to individual species and populations but also to the natural biological communities. It occurs when pollutants are discharged directly or indirectly into water bodies without adequate treatment to remove harmful constituents.

Water pollution is a major problem in the global context. It has been suggested that it is the leading worldwide cause of deaths and diseases, and that it accounts for the deaths of more than 14,000 people daily. In addition to the acute problems of water pollution in developing countries, industrialized countries continue to struggle with pollution problems as well.

Water is typically referred to as polluted when it is impaired by anthropogenic contaminants and either does not support a human use, like serving as drinking water, and/or undergoes a marked shift in its ability to support its constituent biotic communities, such as fish. Natural phenomena such as volcanoes, algae blooms, storms, and earthquakes also cause major changes in water quality and the ecological status of water. Water pollution has many causes and characteristics.

Water Pollution Categories

Surface water and groundwater have often been studied and managed as separate resources, although they are interrelated. Sources of surface water pollution are generally grouped into two categories based on their origin.

Point Source Pollution

Point source pollution refers to contaminants that enter a waterway through a discrete conveyance, such as a pipe or ditch. Examples of sources in this category include discharges from a sewage treatment plant, a factory, or a city storm drain.

Non-point Source Pollution

Non-point source (NPS) pollution refers to diffuse contamination that does not originate from a single discrete source. NPS pollution is often acumulative effect of small amounts of contaminants gathered from a large area. Nutrient runoff in storm water from "sheet flow" over an agricultural field or a forest are sometimes cited as examples of NPS pollution.

Contaminated storm water washed off of parking lots, roads and highways, called urban runoff, is sometimes included under the category of NPS pollution. However, this runoff is typically channeled into storm drain systems and discharged through pipes to local surface waters, and is a point source. The CWA definition of point source was amended in 1987 to include municipal storm sewer systems, as well as industrial storm water, such as from construction sites.

Groundwater Pollution

Interactions between groundwater and surface water are complex. Consequently, groundwater pollution, sometimes referred to as groundwater contamination, is not as easily classified as surface water pollution. By its very nature, groundwater aquifers are susceptible to contamination from sources that may not directly affect surface water bodies, and the distinction of point vs. nonpoint source may be irrelevant. A spill of a chemical contaminant on soil, located away from a surface water body, may not necessarily create point source or non-point source pollution, but nonetheless may contaminate the aquifer below. Analysis of groundwater contamination may focus on soil characteristics and hydrology, as well as the nature of the contaminant itself.

Causes of Water Pollution

The specific contaminants leading to pollution in water include a wide spectrum of chemicals, pathogens, and physical or sensory changes such as elevated temperature and discoloration. While many of the chemicals and substances that are regulated may be naturally occurring (calcium, sodium, iron, manganese, etc.) the concentration is often the key in determining what is a natural component of water, and what is a contaminant.

Oxygen-depleting substances may be natural materials, such as plant matter (e.g. leaves and grass) as well as man-made chemicals. Other natural and anthropogenic substances may cause turbidity (cloudiness) which blocks light and disrupts plant growth, and clogs the gills of some fish species.

Many of the chemical substances are toxic. Pathogens can produce waterborne diseases in either human or animal hosts. Alteration of water's physicai chemistry include acidity (change in pH), electrical conductivity, temperature, and eutrophication. Eutrophication is the fertilization of surface water by nutrients that were previously scarce.

Pathogens

Coliform bacteria are a commonly-used bacterial indicator of water pollution, although not an actual cause of disease. Other microorganisms sometimes found in surface waters which have caused human health problems include:

- Cryptosporidium parvum
- Giardia lamblia
- Salmonella
- Novovirus and other viruses
- Parasitic worms (helminths).

High levels of pathogens may result from inadequately treated sewage discharges. This can be caused by a sewage plant designed with less than secondary treatment (more typical in less-developed countries). In developed countries, older cities with aging infrastructure may have leaky sewage collection systems (pipes, pumps, valves), which can cause sanitary sewer overflows. Some cities also have combined sewers, which may discharge untreated sewage during rain storms.

Pathogen discharges may also be caused by poorly-managed livestock operations.

Chemical and Other Contaminants

Contaminants may include organic and inorganic substances.

Organic water pollutants include:

- Detergents.
- Disinfection by-products found in chemically disinfected drinking water, such as chloroform.
- Food processing waste, which can include oxygen-demanding substances, fats and grease.
- Insecticides and herbicides, a huge range of organohalides and other chemical compounds.
- Petroleum hydrocarbons, including fuels (gasoline, diesel fuel, jet fuels, and fuel oil) and lubricants (motor oil), and fuel combustion byproducts, from stormwater runoff.
- Tree and bush debris from logging operations.
- Volatile organic compounds (VOCs), such as industrial solvents, from improper storage. Chlorinated solvents, which are dense non-aqueous phase liquids (DNAPLs), may fall to the bottom of reservoirs, since they don't mix well with water and are denser.
- Various chemical compounds found in personal hygiene and cosmetic products.

Inorganic Water Pollutants include:

- Acidity caused by industrial discharges (especially sulfur dioxide from power plants)

- Ammonia from food processing waste
- Chemical waste as industrial by-products
- Fertilizers containing nutrients—nitrates and phosphates—which are found in storm water runoff from agriculture, as well as commercial and residential use
- Heavy metals from motor vehicles (via urban storm water runoff) and acid mine drainage
- Silt (sediment) in runoff from construction sites, logging, slash and burn practices or land clearing sites.

Macroscopic pollution—large visible items polluting the water—may be termed "floatables" in an urban storm water context, or marine debris when found on the open seas, and can include such items as:

- Trash (e.g. paper, plastic, or food waste) discarded by people on the ground, and that are washed by rainfall into storm drains and eventually discharged into surface waters
- Nurdles, small ubiquitous waterborne plastic pellets
- Shipwrecks, large derelict ships.

Thermal Pollution

Thermal pollution is the rise or fall in the temperature of a natural body of water caused by human influence. A common cause of thermal pollution is the use of water as a coolant by power plants and industrial manufacturers. Elevated water temperatures decreases oxygen levels (which can kill fish) and affects ecosystem composition, such as invasion by new thermophilic species. Urban runoff may also elevate temperature in surface waters.

Thermal pollution can also be caused by the release of very cold water from the base of reservoirs into warmer rivers.

Transport and Chemical Reactions of Water Pollutants

Most water pollutants are eventually carried by rivers into the oceans. In some areas of the world the influence can be traced hundred miles from the mouth by studies using hydrology transport models. Advanced computer models such as SWMM or the DSSAM Model have been used in many locations worldwide to examine the fate of pollutants in aquatic systems. Indicator filter feeding species such as copepods have also been used to study pollutant fates in the USA. Each successive step up the food chain causes a stepwise concentration of pollutants such as heavy metals (e.g. mercury) and persistent organic pollutants such as DDT. This is known as biomagnification, which is occasionally used interchangeably with bioaccumulation.

Large gyres (vortexes) in the oceans trap floating plastic debris. The North Pacific Gyre for example has collected the so-called "Great Pacific Garbage Patch" that is now estimated at 100 times the size of Texas. Many of these long-lasting pieces wind up in the stomachs of marine birds and animals. This results in obstruction of digestive pathways which leads to reduced appetite or even starvation.

Many chemicals undergo reactive decay or chemically change especially over long periods of time in groundwater reservoirs. A noteworthy class of such chemicals is the chlorinated hydrocarbons such as trichloroethylene (used in industrial metal degreasing and electronics manufacturing) and tetrachloroethylene used in the dry cleaning industry (note latest advances in liquid carbon dioxide in dry cleaning that avoids all use of chemicals). Both of these chemicals, which are carcinogens themselves, undergo partial decomposition reactions, leading to new hazardous chemicals (including dichloroethylene and vinyl chloride).

Groundwater pollution is much more difficult to abate than surface pollution because groundwater can move great distances through unseen aquifers. Non-porous aquifers such as clays partially purify water of bacteria by simple filtration (adsorption and absorption), dilution, and, in some cases, chemical reactions and biological activity: however, in some cases, the pollutants merely transform to soil contaminants. Groundwater that moves through cracks and caverns is not filtered and can be transported as easily as surface water. In fact, this can be aggravated by the human tendency to use natural sinkholes as dumps in areas of Karst topography.

There are a variety of secondary effects stemming not from the original pollutant, but a derivative condition. An example is silt-bearing surface runoff, which can inhibit the penetration of sunlight through the water column, hampering photosynthesis in aquatic plants.

Measurement of Water Pollution

Water pollution may be analyzed through several broad categories of methods: physical, chemical and biological. Most methods involve collection of samples, followed by specialized analytical tests. Some methods may be conducted *in situ,* without sampling, such as temperature. Government agencies and research organizations have published standardized, validated analytical test methods to facilitate the comparability of results from disparate testing events.

Sampling

Sampling of water for physical or chemical testing can be done by several methods, depending on the accuracy needed and the characteristics of the contaminant. Many contamination events are sharply restricted in time, most commonly in association with rain events. For this reason "grab" samples are often inadequate for fully quantifying contaminant levels. Scientists gathering this type of data often employ auto-sampler devices that pump increments of water at either time or discharge intervals.

Sampling for biological testing involves collection of plants and/or animals from the surface water body. Depending on the type of assessment, the organisms may be identified for biosurveys (population counts) and returned to the water body, or they may be dissected for bioassays to determine toxicity.

Chemical Testing

Water samples may be examined using the principles of analytical chemistry. Many published test methods are available for both organic and inorganic compounds. Frequently-used methods include pH, biochemical oxygen demand (BOD), chemical oxygen demand (COD), nutrients (nitrate and phosphorus compounds), metals (including copper, zinc, cadmium, lead and mercury), oil and grease, total petroleum hydrocarbons (TPH), and pesticides.

Control of Water Pollution

Domestic Sewage

In urban areas, domestic sewage is typically treated by centralized sewage treatment plants. Most of these plants are operated by local government agencies. Municipal treatment plants are designed to control conventional pollutants: BOD and suspended solids. Well-designed and operated systems (i.e., secondary treatment or better) can remove 90 per cent or more of these pollutants. Some plants have additional sub-systems to treat nutrients and pathogens. Most municipal plants are not designed to treat toxic pollutants found in industrial wastewater.

Cities with sanitary sewer overflows or combined sewer overflows employ one or more engineering approaches to reduce discharges of untreated sewage, including:

- utilizing a green infrastructure approach to improve storm water management capacity throughout the system

- repair and replacement of leaking and malfunctioning equipment
- increasing overall hydraulic capacity of the sewage collection system (often a very expensive option).

A household or business not served by a municipal treatment plant may have an individual septic tank, which treats the wastewater on site and discharges into the soil. Alternatively, domestic wastewater may be sent to a nearby privately-owned treatment system (e.g. in a rural community).

Industrial Wastewater

Some industrial facilities generate ordinary domestic sewage that can be treated by municipal facilities. Industries that generate wastewater with high concentrations of conventional pollutants (e.g. oil and grease), toxic pollutants (e.g. heavy metals, volatile organic compounds) or other nonconventional pollutants such as ammonia, need specialized treatment systems. Some of these facilities can install a pre-treatment system to remove the toxic components, and then send the partially-treated wastewater to the municipal system. Industries generating large volumes of wastewater typically operate their own complete on-site treatment systems.

Some industries have been successful at redesigning their manufacturing processes to reduce or eliminate pollutants, through a process called pollution prevention.

Heated water generated by power plants or manufacturing plants may be controlled with:

- cooling ponds, man-made bodies of water designed for cooling by evaporation, convection, and radiation
- cooling towers, which transfer waste heat to the atmosphere through evaporation and/or heat transfer
- cogeneration, a process where waste heat is recycled for domestic and/or industrial heating purposes.

9
Ecological Process and Sustainability

Sustainability interfaces with economics through the social and ecological consequences of economic activity. Sustainability economics represents: "... a broad interpretation of ecological economics where environmental and ecological variables and issues are basic but part of a multidimensional perspective. Social, cultural, health-related and monetary/ financial aspects have to be integrated into the analysis." At present the average per capita consumption of people in the developing world is sustainable but population numbers are increasing and individuals are aspiring to high consumption Western lifestyles. The developed world population is only increasing slightly but consumption levels are unsustainable. The challenge for sustainability is to curb and manage Western consumption while raising the standard of living of the developing world without increasing its resource use and environmental impact. This must be done by using strategies and technology that break the link between, on the one hand, economic growth and on the other, environmental damage and resource depletion.

In addressing this issue several key areas have been targeted for economic analysis and reform: the environmental effects of unconstrained economic growth; the consequences of nature being treated as an economic externality; and the possibility of a more ethical economics that takes greater account of the social and environmental consequences of market behaviour.

Decoupling Environmental Degradation and Economic Growth

In the second half of the 20th century world population doubled, food production tripled, energy use quadrupled, and overall economic activity quintupled. Historically there has been a close correlation between economic growth and environmental degradation: as communities grow, so the environment declines. This trend is clearly demonstrated on graphs of human population numbers, economic growth, and environmental indicators. Unsustainable economic growth has been starkly compared to the malignant growth of a cancer because it eats away at the Earth's ecosystem services which are its life-support system. There is concern that, unless resource use is checked, modern global civilization will follow the path of ancient civilizations that collapsed through overexploitation of their resource base. While conventional economics is concerned largely with economic growth and the efficient allocation of resources, ecological economics has the explicit goal of sustainable scale (rather than continual growth), fair distribution and efficient allocation, in that order. The World Business Council for Sustainable Development states that "business cannot succeed in societies that fail". Sustainability studies analyse ways to reduce (decouple) the amount of resource (e.g. water, energy, or materials) needed for the production, consumption and disposal of a unit of good or service whether this be achieved from improved economic management, product design, new technology etc. Ecological economics includes the study of societal metabolism, the throughput of resources that enter and exit the economic system in relation to environmental quality.

Nature as an Economic Externality

The economic importance of nature is indicated by the use of the expression ecosystem services to highlight the market relevance of an increasingly scarce natural world that can no longer be regarded as both unlimited and free. In general as a commodity or service becomes more scarce the price increases and this acts as a restraint that encourages frugality, technical innovation and alternative products. However, this only applies when the product or service falls within the market system. As ecosystem services are generally treated as economic externalities they are unpriced and therefore overused and degraded, a situation sometimes referred to as the Tragedy of the Commons.

One approach to this dilemma has been the attempt to "internalise" these "externalities" by using market strategies like ecotaxes and incentives, tradeable permits for carbon, water and nitrogen use etc., and the encouragement of payment for ecosystem services. Community currencies such as LETS, a gift economy and Time Banking have also been promoted as a way of supporting local economies and the environment. Green economics is another market-based attempt to address issues of equity and the environment.

Economic Opportunity

Treating the environment as an externality may generate short-term profit at the expense of sustainability. Sustainable business practices, on the other hand, integrate ecological concerns with social and economic ones (i.e., the triple bottom line). Growth that depletes ecosystem services is sometimes termed "uneconomic growth" as it leads to a decline in quality of life. Minimizing such growth can provide opportunities for local businesses. For example, industrial waste can be treated as an "economic resource in the wrong place". The benefits of waste reduction include savings from disposal costs, fewer environmental penalties, and reduced liability insurance. This may lead to increased market share due to an improved public image. Energy efficiency can also increase profits by reducing costs.

The idea of sustainability as a business opportunity has led to the formation of organizations such as the Sustainability Consortium of the Society for Organizational Learning, the Sustainable Business Institute, and the World Council for Sustainable Development. The expansion of sustainable business opportunities can contribute to job creation through the introduction of green-collar workers.

Social Dimension

Sustainability issues are generally expressed in scientific and environmental terms, but implementing change is a social challenge that entails, among other things, international and national law, urban planning and transport, local and individual lifestyles and ethical consumerism. "The relationship between human rights and human development, corporate power and environmental justice, global poverty and citizen action, suggest that responsible global citizenship is an inescapable element of what may at first glance seem to be simply matters of personal consumer and moral choice."

Peace, Security, Social Justice

Social disruptions like war, crime and corruption divert resources from areas of greatest human need, damage the capacity of societies to plan for the future, and generally threaten human well-being and the environment. Broad-based strategies for more sustainable social systems include: improved education and the political empowerment of women, especially in developing countries; greater regard for social justice notably equity between rich and poor both within and between countries; and intergenerational equity. Depletion of natural resources including fresh water increases the likelihood of "resource wars". This aspect of sustainability has been referred to as environmental security and creates a clear need for global environmental agreements to manage resources such as aquifers and rivers which span political boundaries, and to protect global systems including oceans and the atmosphere.

Human Settlements

One approach to sustainable living, exemplified by small-scale urban transition towns and rural ecovillages, seeks to create self-reliant communities based on principles of simple living, which maximise self-sufficiency particularly in food production. These principles, on a broader scale, underpin the concept of a bioregional economy. Other approaches, loosely based around new urbanism, are successfully reducing environmental impacts by altering the built environment to create and preserve sustainable cities which support sustainable transport. Residents in compact urban neighbourhoods drive fewer miles, and have significantly lower environmental impacts across a range of measures, compared with those living in sprawling suburbs.

Ultimately, the degree of human progress towards sustainability will depend on large scale social movements which influence both community choices and the built environment. Eco-municipalities may be one such movement. Eco-municipalities take a systems approach, based on sustainability principles. The eco-municipality movement is participatory, involving community members in a bottom-up approach. In Sweden, more than 70 cities and towns — 25 per cent of all municipalities in the country — have adopted a common set of "Sustainability Principles" and implemented these systematically throughout their municipal operations. There are now twelve eco-municipalities in the United States and the American Planning Association has adopted sustainability objectives based on the same principles.

Human Relationship to Nature

According to Murray Bookchin, the idea that humans must dominate nature is common in hierarchical societies.

Bookchin contends that capitalism and market relationships, if unchecked, have the capacity to reduce the planet to a mere resource to be exploited. Nature is thus treated as a commodity: "The plundering of the human spirit by the market place is paralleled by the plundering of the earth by capital." Still more basically, Bookchin argued that most of the activities that consume energy and destroy the environment are senseless because they contribute little to quality of life and well being. The function of work is to legitimse, even create, hierarchy. For this reason understanding the transformation of organic into hierarchical societies is crucial to finding a way forward.

Social ecology, founded by Bookchin, is based on the conviction that nearly all of humanity's present ecological problems originate in, indeed are mere symptoms of, dysfunctional social arrangements. Whereas most authors proceed as if our ecological problems can be fixed by implementing recommendations which stem from physical, biological, economic etc studies, Bookchin's claim is that these problems can only be resolved by understanding the underlying social processes and intervening in those processes by applying the concepts and methods of the social sciences.

Deep ecology establishes principles for the well-being of all life on Earth and the richness and diversity of life forms. This is only compatible with a substantial decrease of the human population and the end of human interference with the nonhuman world. To achieve this, deep ecologists advocate policies for basic economic, technological, and ideological structures that will improve the *quality of life* rather than the *standard of living* . Those who subscribe to these principles are obliged to make the necessary change happen.

Transition

The earth has a finite capacity to provide resources and to absorb waste, and human demands already exceed that capacity. Current lifestyles in the developed world, to which many people in the developing world also aspire, rely on depleting natural capital and are unsustainable. The United Nations has stated, in the Millennium Declaration, that "current unsustainable patterns of production and consumption must be changed". Yet weight of information and scientific evidence is often insufficient to produce necessary social change, especially if that change entails moving people out of their comfort zones.

There is a wealth of advice available to individuals wishing to reduce their personal impact on the environment through small, cheap and easily achievable steps. But the transition required to reduce global human consumption to within sustainable limits involves much larger changes, at all levels and contexts of society. The United Nations have recognised the central role of education, and have declared a decade of education for sustainable development, 2005-2014, which aims to "challenge us all to adopt new behaviours and practices to secure our future". The Worldwide Fund for Nature proposes a strategy for sustainability that goes beyond education to tackle underlying individualistic and materialistic societal values head-on and strengthen people's connections with the natural world. The level of change required to safeguard the life-supporting capacity of the Earth sets new challenges for community and political structures.

SUSTAINABILITY

Sustainability, in a broad sense, is the capacity to endure. It can be defined as the ability of an ecosystem to maintain ecological processes, functions, biodiversity and productivity into the future. In ecology, the word describes how biological systems remain diverse and productive over time. For humans it is the potential for long-term maintenance of wellbeing, which in turn depends on the wellbeing of the natural world and the responsible use of natural resources.

The word 'sustainability' has become a wide-ranging term that can be applied to almost every facet of life on Earth, from a local to a global scale and over various time periods. Long-lived and healthy wetlands and forests are examples of sustainable biological systems. Invisible chemical cycles redistribute water, oxygen, nitrogen and carbon through the world's living and non-living systems, and have sustained life for millions of years. As the earth's human population has increased, natural ecosystems have declined and changes in the balance of natural cycles has had a negative impact on both humans and other living systems.

There is now abundant scientific evidence that humanity is living unsustainably. Returning human use of natural resources to within sustainable limits will require a major collective effort. Since the 1980s, human sustainability has implied the integration of economic, social and environmental spheres to: "meet the needs of the present without compromising the ability of future generations to meet their own needs."

Efforts to live more sustainably can take many forms from reorganizing living conditions (e.g., ecovillages, eco-municipalities and sustainable cities), reappraising economic sectors (green building, sustainable agriculture), or work practices (sustainable architecture), using science to develop new technologies (green technologies, renewable energy), to adjustments in individual lifestyles.

Although the definition of sustainable development (above), given by the Brundtland Commission, is frequently quoted, it is not universally accepted and has undergone various interpretations. Definitions of sustainability may be expressed as statements of fact, intent, or value with sustainability treated as either a "journey" or "destination". Where we are now, where we need to be going, and how we are to get there are all open to interpretation and will depend on the particular context under consideration. What can meaningfully be described as sustainable will depend on the scale of space and time that is appropriate to the item under consideration. For example, if time criteria have not been met, then assertions of sustainability are more like predictions than definitions. This difficult mix has been described as a "dialogue of values that defies consensual definition." Sustainability has been regarded as both an important but unfocused concept like "liberty" or "justice" and as a feel-good buzzword with little meaning or substance. The idea of sustainable development is sometimes viewed as an oxymoron because development inevitably depletes and degrades the environment. Consequently some definitions either avoid the word *development* and use the term *sustainability* exclusively, or emphasise the environmental component, as in "environmentally sustainable development."

The dimensions of sustainability are often taken to be: environmental, social and economic, known as the "three pillars". These can be depicted as three overlapping circles (or ellipses), to show that they are not mutually exclusive and can be mutually reinforcing. While this model initially improved the standing of environmental concerns, it has since been criticised for not adequately showing that societies and economies are fundamentally reliant on the natural world. According to English environmentalist and author Jonathon Porritt, "The economy is, in the first instance, a subsystem of human society ... which is itself, in the second instance, a subsystem of the totality of life on Earth (the biosphere). And no subsystem can expand beyond the capacity of the total system of which it is a part." For this reason a second diagram shows economy as a component of society, both bounded by, and dependent upon, the environment. As the American World Bank ecological economist Herman Daly famously asked, "what use is a sawmill without a forest?" The concept of living within environmental constraints yields another definition of sustainability: "improving the quality of human life while living within the carrying capacity of supporting eco-systems."

The Earth Charter goes beyond defining what sustainability is, and seeks to establish the values and direction needed to achieve it: "We must join together to bring forth a sustainable global society founded on respect for nature, universal human rights, economic justice, and a culture of peace. Towards this end, it is imperative that we, the peoples of Earth, declare our responsibility to one another, to the greater community of life, and to future generations."

In early human history the environmental impacts of small bands of hunter-gatherers would have been relatively small, even though the use of fire and the desire for specific foods may have altered the natural composition of plant and animal communities. With the Neolithic Revolution 8,000 and 10,000 years ago came the emergence of agriculture and settled communities. Societies outgrowing their local food supply or depleting critical resources either moved on or faced collapse. In contrast, stable communities of shifting cultivators and horticulturists existed in New Guinea and South America, and large agrarian communities in China, India, Polynesia and elsewhere have farmed in the same localities for centuries.

Technological advances over several millennia gave humans increasing control over the environment. But it was the Western industrial revolution of the 17th to 19th centuries that tapped into the vast growth potential of the energy in fossil fuels to power sophisticated machinery technology. Such conditions led to a human population explosion and unprecedented industrial, technological and scientific growth that has continued to this day. From 1650 to 1850 the global population doubled from around 500 million to 1 billion people.

By the 20th century, the industrial revolution had resulted in an exponential increase in the human consumption of resources and an increase in health, wealth and population. Ecology as a new scientific discipline was gaining general acceptance and ideas now part and parcel to sustainability were being explored including the recognition of the interconnectedness of living systems, the importance of global natural cycles, the passage of energy through trophic levels of living systems.

After the deprivations of the great depression and World War II the developed world entered a post-1950s "great acceleration" of growth and population while a gathering environmental movement pointed out that there were environmental costs associated with the many material benefits that were now being enjoyed. Technological innovations included plastics, synthetic chemicals and nuclear energy as fossil fuels also continued to transform society. The negative influences of the new technology were documented by American marine biologist and naturalist Rachel Carson in her influential book *Silent Spring* in 1962. A period of peak oil production was anticipated in 1956 by American geoscientist M. King Hubbert's peak oil. In the 1970s environmentalism's concern with pollution, the population explosion, consumerism and the depletion of finite resources found expression in *Small Is Beautiful,* by British economist E.F. Schumacher in 1973, and *The Limits to Growth* published by the global think tank, the Club of Rome, in 1975.

By the late 20th century environmental problems were becoming global in scale. and the 1973 and 1979 energy crises demonstrated the extent to which the global community had become dependent on a nonrenewable resource.

In 1987, the United Nation's World Commission on Environment and Development (the Brundtland Commission), in its report *Our Common Future* suggested that sustainable development was needed to meet human needs while not increasing environmental problems. In 1961, almost all countries in the world had the capacity to meet their own demand but by 2005 the situation had changed and many countries were able to meet their needs only by importing resources from other nations. A move toward more sustainable living emerged, based on increasing public awareness and adoption of recycling, and renewable energies. The development of renewable sources of energy in the 1970s and 80's, primarily in wind turbines and photovoltaics and increased use of hydroelectricity, presented more sustainable alternatives to fossil fuel and nuclear energy generation.

In the 21st century there is heightened awareness of the threat posed by the human-induced greenhouse effect. Ecological economics now seeks to bridge the gap between ecology and traditional neoclassical economics: and proposes an inclusive and ethical economic model for society. Many new techniques have arisen to help measure and implement sustainability, including Life Cycle Assessment to Cradle, the Ecological Footprint Analysis, and green building. The work of Bina Agarwal and Vandana Shiva amongst many others, has brought some of the cultural knowledge of traditional, sustainable agrarian societies into the academic discourse on sustainability, and blended that knowledge with modern scientific principles.

Principles and Concepts

Sustainability science and environmental science form the basis for much of the philosophical and analytic framework of sustainability. Quantitative data is collected through sustainability measurement and this data is then used in governance for sustainability.

Scale and Context

Sustainability is studied and managed over many scales (levels or frames of reference) of time and space and in many contexts of environmental, social and economic organization. The focus ranges from the total carrying capacity (sustainability) of planet Earth to the sustainability of economic sectors, ecosystems, countries, municipalities, neighbourhoods, home gardens, individual lives, individual goods and services, occupations, lifestyles, behaviour patterns and so on. In a smaller form, it can entail the full compass of biological and human activity or any part of it. As Daniel Botkin, author and environmentalist, has stated: "We see a landscape that is always in flux, changing over many scales of time and space."

Global Goals

At the global level a number of key goals have been isolated:

- *Intergenerational equity* — providing future generations with the same environmental potential as presently exists
- *Decoupling economic growth from environmental degradation* — managing economic growth to be less resource intensive and less polluting
- *Integration of all pillars* — integrating environmental, social and economic sectors when developing sustainability policies
- *Ensuring environmental adaptability and resilience* — maintaining and enhancing the adaptive capacity of the environmental system
- Preventing irreversible long-term damage to ecosystems and human health

- *Ensuring distributional equity* — avoiding unfair or high environmental costs on vulnerable populations
- *Accepting global responsibility* - assuming responsibility for environmental effects that occur outside areas of jurisdiction
- *Education and grassroots involvement* - people and communities investigating problems and developing new solutions.

Consumption, Population, Technology, Resources

The overall driver of human impact on Earth systems is the consumption of biophysical resources. Human consumption can be divided into three key components: population numbers, levels of consumption (affluence), and impact per unit of resource use (which depends on the technology used). This has been expressed through an equation:

$$I = P \times A \times T$$

Where: I = Environmental impact, P = Population,
A = Affluence, T = Technology

Historically, humanity has responded to a demand for more resources by trying to increase supply. Sustainability, instead, applies demand management of all goods and services by promoting reduced consumption, using renewable resources where possible, and encouraging practices that minimise resource intensity while maximising resource productivity. Careful resource management is applied at many scales, but especially at the levels of economic sectors like agriculture, manufacturing and industry as well as to individual goods and services and the consumption patterns of households and individuals.

Population

According to the 2008 Revision of the official United Nations population estimates and projections, the world population is projected to reach 7 billion early in 2012, up from the current 6.9 billion (May 2009), to exceed 9 billion people by 2050. Most of the increase will be in developing countries whose population is projected to rise from 5.6 billion in 2009 to 7.9 billion in 2050. This increase will be distributed among the population aged 15–59 (1.2 billion) and 60 or over (1.1 billion) because the number of children under age 15 in developing countries will decrease. In contrast, the population of the more developed regions is expected to undergo only slight increase from 1.23 billion to 1.28 billion, and this would have declined to 1.15 billion but for a projected net migration from developing to developed countries, which is expected to average 2.4 million persons annually from 2009 to 2050. Long-term estimates of global population suggest a peak at around 2070 of nine to ten billion people, and then a slow decrease to 8.4 billion by 2100.

Emerging economies like those of China and India aspire to the living standards of the Western world as does the non-industrialized world in general. It is the combination of population increase in the developing world and unsustainable consumption levels in the developed world that poses a stark challenge to sustainability.

Direct and Indirect Environmental Impacts

At a fundamental level energy flow and biogeochemical cycling set an upper limit on the number and mass of organisms in any ecosystem. Human impacts on the Earth are demonstrated through detrimental changes in the global biogeochemical cycles of chemicals that are critical to life, most notably those of water, oxygen, carbon, nitrogen and phosphorus.

Measurement

To survive on planet Earth humans must live within its measurable biophysical constraints. By establishing quantitative measures for sustainability it is possible to set goals, apply management strategies, and measure progress. The Natural Step (TNS) framework developed by Karl-Henrik Robèrt examines sustainability and resource use from its thermodynamic foundations to determine how humans use and apportion natural capital in a way that is sustainable and just. The TNS framework's system conditions of sustainability provide a means for the scientifically-based measurement of sustainability. Natural capital includes resources from the earth's crust (i.e., minerals, oil), those produced by humans (synthetic substances), and those of the biosphere. Equitable access to natural capital is also a component of sustainability. The energy generated in use of resources—referred to as exergy—can be measured as the embodied energy of a product or service over its life cycle. Its analysis, using methods such as Life Cycle Analysis or Ecological Footprint analysis provide basic indicators of sustainability on various scales.

There is now a vast number of sustainability indicators, metrics, benchmarks, indices, reporting procedures, audits and more. They include environmental, social and economic measures separately or together over many scales and contexts. Environmental factors are integrated with economics through ecological economics, resource economics and thermoeconomics, and social factors through metrics like the Happy Planet Index which measures the well-being of people in the nations of the world while taking into account their environmental impact. Some of the best known and most widely used sustainability measures are listed in the side bar, they include corporate sustainability reporting, Triple Bottom Line accounting, and estimates of the quality of sustainability governance for individual countries using the Environmental Sustainability Index and Environmental Performance Index.

Carrying Capacity

Is humanity living within the carrying capacity of the planet – are humans living sustainably on planet Earth? The

Ecological footprint measures human consumption in terms of the biologically productive land needed to provide the resources, and absorb the wastes of the average global citizen. In 2008 it required 2.7 global hectares per person, 30% more than the natural biological capacity of 2.1 global hectares (assuming no provision for other organisms). The resulting ecological deficit must be met from unsustainable *extra* sources and these are obtained in three ways: embedded in the goods and services of world trade; taken from the past (e.g. fossil fuels); or borrowed from the future as unsustainable resource usage (e.g. by over exploiting forests and fisheries).

The figure (right) indicates the sustainability of a range of countries in terms of the Ecological Footprint compared to the UN Human Development Index (a measure of standard of living): it shows what is necessary for countries to maintain an acceptable standard of living for their citizens while, at the same time, living at a globally sustainable level. The general trend is for higher standards of living to become less sustainable. As always population growth has a marked influence on levels of consumption and the efficiency of resource use. At present Cuba is the best example in this category. The sustainability goal is to raise the global standard of living without increasing the use of resources beyond globally sustainable levels; that is, to not exceed "one planet" consumption. A wealth of information generated by reports at the national, regional and city scales confirm the global trend towards societies that are becoming less sustainable over time.

Global Human Impact on Biodiversity

The Millennium Ecosystem Assessment is an international synthesis by over 1000 of the world's leading biological scientists that analyses the state of the Earth's ecosystems. It concludes that human activity is having a significant and escalating impact on the biodiversity of world ecosystems, reducing both their resilience and biocapacity. The report refers to natural systems as humanity's "life-support system", providing essential "ecosystem services". The assessment measures 24 ecosystem services concluding that only four have shown improvement over the last 50 years, 15 are in serious decline, and five are in a precarious condition.

Environmental Dimension

Healthy ecosystems provide vital goods and services to humans and other organisms. There are two major ways of reducing negative human impact and enhancing ecosystem services:

(a) Environmental management. This direct approach is based largely on information gained from earth science, environmental science and conservation biology.

However, this is management at the end of a long series of indirect causal factors that are initiated by human consumption, so a second approach is through demand management of human resource use.

(b) Management of human consumption of resources, an indirect approach based largely on information gained from economics. Herman Daly has suggested three broad criteria for ecological sustainability: renewable resources should provide a sustainable yield (the rate of harvest should not exceed the rate of regeneration); for non-renewable resources there should be equivalent development of renewable substitutes; waste generation should not exceed the assimilative capacity of the environment.

ENVIRONMENTAL MANAGEMENT

At the global scale and in the broadest sense environmental management involves the oceans, freshwater systems, land and atmosphere, but following the sustainability principle of scale it can be equally applied to any ecosystem from a tropical rainforest to a home garden.

Atmosphere

In March 2009 at a meeting of the Copenhagen Climate Council 2,500 climate experts from 80 countries issued a keynote statement that there is now "no excuse" for failing to act on global warming and that without strong carbon reduction targets "abrupt or irreversible" shifts in climate may occur that "will be very difficult for contemporary societies to cope with". Management of the global atmosphere now involves assessment of all aspects of the carbon cycle to identify opportunities to address human-induced climate change and this has become a major focus of scientific research because of the potential catastrophic effects on biodiversity and human communities.

Other human impacts on the atmosphere include the air pollution in cities, the pollutants including toxic chemicals like nitrogen oxides, sulphur oxides, volatile organic compounds and particulate matter that produce photochemical smog and acid rain, and the chlorofluorocarbons that degrade the ozone layer. Anthropogenic particulates such as sulphate aerosols in the atmosphere reduce the direct irradiance and reflectance (albedo) of the Earth's surface. Known as global dimming the decrease is estimated to have been about 4% between 1960 and 1990 although the trend has subsequently reversed.

Global dimming may have disturbed the global water cycle by reducing evaporation and rainfall in some areas. It also creates a cooling effect and this may have partially masked the effect of greenhouse gases on global warming.

Oceans

Ocean circulation patterns have a strong influence on climate and weather and, in turn, the food supply of both humans and other organisms. Scientists have warned of the possibility, under the influence of climate change, of a sudden alteration in circulation patterns of ocean currents that could drastically alter the climate in some regions of the globe. Major human environmental impacts occur in the more habitable regions of the ocean fringes – the estuaries, coastline and bays. Ten per cent of the world's population – about 600 million people – live in low-lying areas vulnerable to sea level rise. Trends of concern that require management include: over-fishing (beyond sustainable levels); coral bleaching due to ocean warming and ocean acidification due to increasing levels of dissolved carbon dioxide; and sea level rise due to climate change. Because of their vastness oceans also act as a convenient dumping ground for human waste. Remedial strategies include: more careful waste management, statutory control of overfishing by adoption of sustainable fishing practices and the use of environmentally sensitive and sustainable aquaculture and fish farming, reduction of fossil fuel emissions and restoration of coastal and other marine habitat.

Freshwater

Water covers 71% of the Earth's surface. Of this, 97.5% is the salty water of the oceans and only 2.5% freshwater, most of which is locked up in the Antarctic ice sheet. The remaining freshwater is found in lakes, rivers, wetlands, the soil, aquifers and atmosphere. All life depends on the solar-powered global water cycle, the evaporation from oceans and land to form water vapour that later condenses from clouds as rain, which then becomes the renewable part of the freshwater supply. Awareness of the global importance of preserving water for ecosystem services has only recently emerged as, during the 20th century, more than half the world's wetlands have been lost along with their valuable environmental services. Biodiversity-rich freshwater ecosystems are currently declining faster than marine or land ecosystems making them the world's most vulnerable habitats. Increasing urbanization pollutes clean water supplies and much of the world still does not have access to clean, safe water. In the industrial world demand management has slowed absolute usage rates but increasingly water is being transported over vast distances from water-rich natural areas to population-dense urban areas and energy-hungry desalination is becoming more widely used. Greater emphasis is now being placed on the improved management of blue (harvestable) and green (soil water available for plant use) water, and this applies at all scales of water management.

Land

Loss of biodiversity stems largely from the habitat loss and fragmentation produced by the human appropriation of land for development, forestry and agriculture as natural capital is progressively converted to man-made capital. Land use change is fundamental to the operations of the biosphere because alterations in the relative proportions of land dedicated to urbanization, agriculture, forest, woodland, grassland and pasture have a marked effect on the global water, carbon and nitrogen biogeochemical cycles and this can impact negatively on both natural and human systems. At the local human scale major sustainability benefits accrue from the pursuit of green cities and sustainable parks and gardens.

Forests

Since the Neolithic Revolution about 47% of the world's forests have been lost to human use. Present-day forests occupy about a quarter of the world's ice-free land with about half of these occurring in the tropics In temperate and boreal regions forest area is gradually increasing (with the exception of Siberia), but deforestation in the tropics is of major concern.

Forests moderate the local climate and the global water cycle through their light reflectance (albedo) and evapotranspiration. They also conserve biodiversity, protect water quality, preserve soil and soil quality, provide fuel and pharmaceuticals, and purify the air. These free ecosystem services have no market value and so forest conservation has little appeal when compared with the economic benefits of logging and clearance which, through soil degradation and organic decomposition returns carbon dioxide to the atmosphere. The United Nations Food and Agriculture Organization (FAO) estimates that about 90% of the carbon stored in land vegetation is locked up in trees and that they sequester about 50% more carbon than is present in the atmosphere. Changes in land use currently contribute about 20% of total global carbon emissions (heavily logged Indonesia and Brazil are a major source of emissions). Climate change can be mitigated by sequestering carbon in reafforestation schemes, plantations and timber products. Also wood biomass can be utilized as a renewable carbon-neutral fuel. The FAO has suggested that, over the period 2005–2050, effective use of tree planting could absorb about 10–20% of man-made emissions – so monitoring the condition of the world's forests must be part of a global strategy to mitigate emissions and protect ecosystem services. However, climate change may pre-empt this FAO scenario as a study by the International Union of Forest Research Organizations in 2009 concluded that the stress of a 2.5C (4.5F) temperature rise above pre-industrial levels could result in the release of vast amounts of carbon so the potential of forests to act as carbon "sinks" is "at risk of being lost entirely".

Cultivated Land

Feeding more than six billion human bodies takes a heavy toll on the Earth's resources. This begins with the appropriation of about 38% of the Earth's land surface and about 20% of its net primary productivity. Added to this are the resource-hungry activities of industrial agribusiness – everything from the crop need for irrigation water, synthetic fertilizers and pesticides to the resource costs of food packaging, transport (now a major part of global trade) and retail. Food is essential to life. But the list of environmental costs of food production is a long one: topsoil depletion, erosion and conversion to desert from constant tillage of annual crops; overgrazing; salinization; sodification; waterlogging; high levels of fossil fuel use; reliance on inorganic fertilizers and synthetic organic pesticides; reductions in genetic diversity by the mass use of monocultures; water resource depletion; pollution of waterbodies by run-off and groundwater contamination; social problems including the decline of family farms and weakening of rural communities.

All of these environmental problems associated with industrial agriculture and agribusiness are now being addressed through such movements as sustainable agriculture, organic farming and more sustainable business practices.

Extinctions

Although biodiversity loss can be monitored simply as loss of species, effective conservation demands the protection of species within their natural habitats and ecosystems. Following human migration and population growth, species extinctions have progressively increased to a rate unprecedented since the Cretaceous–Tertiary extinction event. Known as the Holocene extinction event this current human-induced extinction of species ranks as one of the worlds six mass extinction events. Some scientific estimates indicate that up to half of presently existing species may become extinct by 2100. Current extinction rates are 100 to 1000 times their prehuman levels with more than 10% birds and mammals threatened, about 8% of plants, 5% of fish and more than 20% of freshwater species.

The 2008 IUCN Red List warns that long-term droughts and extreme weather put additional stress on key habitats and, for example, lists 1,226 bird species as threatened with extinction, which is one-in-eight of all bird species. The Red List Index also identifies 44 tree species in Central Asia as under threat of extinction due to over-exploitation and human development and threatening the region's forests which are home to more than 300 wild ancestors of modern domesticated fruit and nut cultivars.

Biological Invasions

In many parts of the industrial world land clearing for agriculture has diminished and here the greatest threat to biodiversity, after climate change, has become the destructive effect of invasive species. Increasingly efficient global transport has facilitated the spread of organisms across the planet. The potential danger of this aspect of globalization is starkly illustrated through the spread of human diseases like HIV AIDS, mad cow disease, bird flu and swine flu, but invasive plants and animals are also having a devastating impact on native biodiversity. Non-indigenous organisms can quickly occupy disturbed land and natural areas where, in the absence of their natural predators, they are able to thrive. At the global scale this issue is being addressed through the Global Invasive Species Information Network but there is improved international biosecurity legislation to minimize the transmission of pathogens and invasive organisms. Also, through CITES legislation there is control the trade in rare and threatened species. Increasingly at the local level public awareness programmes are alerting communities, gardeners, the nursery industry, collectors, and the pet and aquarium industries, to the harmful effects of potentially invasive species.

MANAGEMENT OF HUMAN CONSUMPTION

The underlying driver of direct human impacts on the environment is human consumption. This impact is reduced by not only consuming less but by also making the full cycle of production, use and disposal more sustainable. Consumption of goods and services can be analysed and managed at all scales through the chain of consumption, starting with the effects of individual lifestyle choices and spending patterns, through to the resource demands of specific goods and services, the impacts of economic sectors, through national economies to the global economy. Analysis of consumption patterns relates resource use to the environmental, social and economic impacts at the scale or context under investigation. The ideas of embodied resource use (the total resources needed to produce a product or service), resource intensity (the resources needed for each dollar spent on a good or service), and resource productivity (the amount of good or service produced for a given input of resource) are important tools for understanding the impacts of consumption with simple key resource categories indicating human needs being food, energy, materials and water.

Energy

The Sun's energy, stored by plants (primary producers) during photosynthesis, passes through the food chain to other organisms to ultimately power all living processes. Since the industrial revolution the concentrated energy of the Sun stored in fossilized plants as fossil fuels has been a major driver of technology which, in turn, has been the source of both economic and political power. In 2007 climate scientists of the IPCC concluded that there was at least a 90% probability that atmospheric increase in CO_2 was human-induced, mostly as a result of fossil fuel emissions but, to a lesser extent from changes in land use. Stabilize the world's climate will require high income countries to reduce their emissions by 60-90% over 2006 levels by 2050 which should hold CO_2 levels at 450-650 ppm from current levels of about 380 ppm. Above this level and temperatures could rise by more than 2°C (36°F) to produce "catastrophic" climate change. Reduction of current

CO_2 levels must be achieved against a background of global population increase and developing countries aspiring to energy-intensive high consumption Western lifestyles.

Reducing greenhouse emissions, referred to as decarbonization, is being tackled at all scales, ranging from tracking the passage of carbon through the carbon cycle to the exploration of renewable energies, developing less carbon-hungry technology and transport systems and attempts by individuals to lead carbon neutral lifestyles by monitoring the fossil fuel use embodied in all the goods and services they use.

Water

Water security and food security are inextricably linked. In the decade 1951-60 human water withdrawals were four times greater than the previous decade. This rapid increase resulted from scientific and technological developments impacting through the economy — especially the increase in irrigated land, growth in industrial and power sectors, and intensive dam construction on all continents. This altered the water cycle of rivers and lakes, affected their water quality and had a significant impact on the global water cycle. Currently towards 35% of human water use is unsustainable, drawing on diminishing aquifers and reducing the flows of major rivers: this percentage is likely to increase if climate change worsens, populations increase, aquifers become progressively depleted and supplies become polluted and unsanitary. From 1961 to 2001 water demand doubled - agricultural use increased by 75%, industrial use by more than 200%, and domestic use more than 400%. Humans currently use 40-50% of the globally available freshwater in the approximate proportion of 70% for agriculture, 22% for industry, and 8% for domestic purposes and the total volume is progressively increasing.

Water efficiency is being improved on a global scale by increased demand management, improved infrastructure, improved water productivity of agriculture, minimising the water intensity (embodied water) of goods and services, addressing shortages in the non-industrialised world, concentrating food production in areas of high productivity; and planning for climate change. At the local level people are becoming more water-self-sufficient by harvesting rainwater and reducing use of mains water.

Food

According to American Public Health Association (APHA) a "sustainable food system" as "one that provides healthy food to meet current food needs while maintaining healthy ecosystems that can also provide food for generations to come with minimal negative impact to the environment. A sustainable food system also encourages local production and distribution infrastructures and makes nutritious food available, accessible, and affordable to all. Further, it is humane and just, protecting farmers and other workers, consumers, and communities." Concerns about the environmental impacts of agribusiness and the stark contrast between the obesity problems of the Western world and the poverty and food insecurity of the developing world have generated a strong movement towards healthy, sustainable eating as a major component of overall ethical consumerism. The environmental effects of different dietary patterns depend on many factors, including the proportion of animal and plant foods consumed and the method of food production. The World Health Organization has published a *Global Strategy on Diet, Physical Activity and Health* which was endorsed by the May 2004 World Health Assembly. It recommends the Mediterranean diet which is associated with health and longevity and is low in meat, rich in fruits and vegetables, low in added sugar and limited salt, and low in saturated fatty acids; the traditional source of fat in the Mediterranean is olive oil, rich in monounsaturated fat. The healthy rice-based Japanese diet is also high in carbohydrates and low in fat. Both diets are low in meat and saturated fats and high in legumes and other vegetables; they are associated with a low incidence of ailments and low environmental impact.

At the global level the environmental impact of agribusiness is being addressed through sustainable agriculture and organic farming. At the local level there are various movements working towards local food production, more productive use of urban wastelands and domestic gardens including permaculture, urban horticulture, local food, slow food, sustainable gardening, and organic gardening.

As global population and affluence increases, so does the use of materials which has increased in volume, diversity and distance transported. Included here are raw materials, minerals, synthetic chemicals (including hazardous substances), manufactured products, food, living organisms and waste.

Materials

Sustainable use of materials has targeted the idea of dematerialization, converting the linear path of materials (extraction, use, disposal in landfill) to a circular material flow that reuses materials as much as possible, much like the cycling and reuse of waste in nature. This approach is supported by product stewardship and the increasing use of material flow analysis at all levels, especially individual countries and the global economy.

Toxic Substances

Synthetic chemical production has escalated following the stimulus it received during the second World War. Chemical production includes everything from herbicides, pesticides and fertilizers to domestic chemicals and hazardous substances. Apart from the build-up of greenhouse gas emissions in the atmosphere, chemicals of particular concern include: heavy metals, nuclear waste, chlorofluorocarbons, persistent organic pollutants and all harmful chemicals capable of bioaccumulation. Although most synthetic chemicals are harmless there needs to be rigorous testing of new chemicals, in all countries, for adverse environmental and health effects. International legislation has been established to deal with the global distribution and management of dangerous goods.

Waste

Every economic activity produces material that can be classified as waste. The average human uses 45-85 tonnes of materials each year. To reduce waste industry, business and government are now mimicking nature by turning the waste produced by industrial metabolism into resource. Dematerialization is being encouraged through the ideas of industrial ecology, ecodesign and ecolabelling. In addition to the well-established "reduce, reuse and recycle" shoppers are using their purchasing power for ethical consumerism.

10
Environmentalism and Pollution Prevention

Environmentalism can also be defined as a social movement that seeks to influence the political process by lobbying, activism, and education in order to protect natural resources and ecosystems. In recognition of humanity as a participant in ecosystems, the environmental movement is centered on ecology, health, and human rights.

An environmentalist is a person who may advocate the sustainable management of resources and stewardship of the natural environment through changes in public policy or individual behaviour. In various ways (for example, grassroots activism and protests), environmentalists and environmental organizations seek to give the natural world a stronger voice in human affairs.

Though opinions vary, environmentalism may be seen as a spectrum; from the reformist to the radical.

A concern for environmental protection has recurred in diverse forms, in different parts of the world, throughout history. For example, in the Middle East, the earliest known writings concerned with environmental pollution were Arabic medical treatises written during the "Arab Agricultural Revolution", by writers such as Alkindus, Costa ben Luca, Rhazes, Ibn Al-Jazzar, al-Tamimi, al-Masihi, Avicenna, Ali ibn Ridwan, Isaac Israeli ben Solomon, Abd-el-latif, and Ibn al-Nafis. They were concerned with air contamination, water contamination, soil contamination, solid waste mishandling, and environmental assessments of certain localities.

In Europe, King Edward I of England banned the burning of sea-coal by proclamation in London in 1272, after its smoke had become a problem. But the fuel was so common in England that this earliest of names for it was acquired because it could be carted away from some shores by the wheelbarrow. Air pollution would continue to be a problem there, especially later during the industrial revolution, and extending into the recent past with the Great Smog of 1952.

ORIGINS OF ENVIRONMENTAL MOVEMENT

In Europe, it was the Industrial Revolution that gave rise to modern environmental pollution as it is generally understood today. The emergence of great factories and consumption of immense quantities of coal and other fossil fuels gave rise to unprecedented air pollution and the large volume of industrial chemical discharges added to the growing load of untreated human waste. The first large-scale, modern environmental laws came in the form of the British Alkali Acts, passed in 1863, to regulate the deleterious air pollution (gaseous hydrochloric acid) given off by the Leblanc process, used to produce soda ash. Environmentalism grew out of the amenity movement, which was a reaction to industrialization, the growth of cities, and worsening air and water pollution.

In the United States, the beginnings of an environmental movement can be traced as far back as 1739, when Benjamin Franklin and other Philadelphia residents, citing "public rights," petitioned the Pennsylvania Assembly to stop waste dumping and remove tanneries from Philadelphia's commercial district. The US movement expanded in the 1800s, out of concerns for protecting the natural resources of the West, with individuals such as John Muir and Henry David Thoreau making key philosophical contributions. Thoreau was interested in peoples' relationship with nature and studied this by living close to nature in a simple life. He published his experiences in the book *Walden,* which argues that people should become intimately close with nature. Muir came to believe in nature's inherent right, especially after spending time hiking in Yosemite Valley and studying both the ecology and geology. He successfully lobbied Congress to form Yosemite National Park and went on to set up the Sierra Club. The conservationist principles as well as the belief in an inherent right of nature were to become the bedrock of modern environmentalism.

In the 20th century, environmental ideas continued to grow in popularity and recognition. Efforts were starting to be made to save some wildlife, particularly the American Bison. The death of the last Passenger Pigeon as well as the endangerment of the American Bison helped to focus the minds of conservationists and popularize their concerns. Notably in 1916 the National Park Service was founded by President Woodrow Wilson.

In the 1970s, the Chipko movement was formed in India; influenced by Mahatma Gandhi, they setup peaceful resistance to deforestation by literally hugging trees (leading to the term "tree huggers"). Their peaceful methods of protest and slogan "ecology is permanent economy" were very influential.

By the mid-1970s, many felt that people were on the edge of environmental catastrophe. The Back-to-the-land movement started to form and ideas of environmental ethics joined with anti-Vietnam War sentiments and other political issues. These individuals lived outside normal society and started to take on some of the more radical environmental theories such as deep ecology. Around this time more mainstream environmentalism was starting to show force with the signing of the Endangered Species Act in 1973 and the formation of CITES in 1975.

Environmentalism has also changed to deal with new issues such as global warming and genetic engineering.

Environmental Movement

The environmental movement (a term that sometimes includes the conservation and green movements) is a diverse scientific, social, and political movement. In general terms, environmentalists advocate the sustainable management of resources, and the protection (and restoration, when necessary) of the natural environment through changes in public policy and individual behavior. In its recognition of humanity as a participant in ecosystems, the movement is centered around ecology, health, and human rights. Additionally, throughout history, the movement has been incorporated into religion. The movement is represented by a range of organizations, from the large to grassroots, but a younger demographic than is common in other social movements. Because of its large membership, varying and strong beliefs, the movement is not entirely united. Indeed, some argue that an environmental ethic of at least some sort is so urgently needed in all quarters that the broader the better. Conversely, disunity can be a weakness in the face of strong opposition from unsympathetic political and industrial forces.

Free Market Environmentalism

Free market environmentalism is a theory that argues that the free market, property rights, and tort law provide the best tools to preserve the health and sustainability of the environment. It considers environmental stewardship to be natural, as well as the expulsion of pollutors and other aggressors through individual and class action.

Preservation and Conservation

Environmental preservation in the United States is viewed as the setting aside of natural resources to prevent damage caused by contact with humans or by certain human activities, such as logging, mining, hunting, and fishing, only to replace them with new human activities such as tourism and recreation. Regulations and laws may be enacted for the preservation of natural resources.

Dark Greens, Light Greens, and Bright Greens

Contemporary environmentalists are often described as being split into three groups: *Dark, Light,* and *Bright Greens.*

Light Greens see protecting the environment first and foremost as a personal responsibility. They fall in on the reformist end of the spectrum introduced above, but light Greens do not emphasize environmentalism as a distinct political ideology, or even seek fundamental political reform. Instead they often focus on environmentalism as a lifestyle choice. The motto "Green is the new black." sums up this way of thinking, for many.

In contrast, Dark Greens believe that environmental problems are an inherent part of industrialized civilization evident in both state socialist and capitalist societies, and seek radical political change. 'Dark greens' tend to believe that dominant political ideologies (sometimes referred to as industrialism) are corrupt and inevitably lead to consumerism, alienation from nature and resource depletion. Dark Greens claim that this is caused by the emphasis on growth that exists within all existing ideologies, a tendency referred to as 'growth mania'. The dark green brand of environmentalism is associated with ideas of Deep Ecology, Post-materialism, Holism, the Gaia Theory of James Lovelock and the work of Fritjof Capra.

More recently, a third group may be said to have emerged in the form of Bright Greens. This group believes that radical changes are needed in the economic and political operation of society in order to make it sustainable, but that better designs, new technologies and more widely distributed social innovations are the means to make those changes— and that we can neither shop nor protest our way to sustainability. As Ross Robertson writes, "right green environmentalism is less about the problems and limitations we need to overcome than the "tools, models, and ideas" that already exist for overcoming them. It forgoes the bleakness of protest and dissent for the energizing confidence of constructive solutions."

Environmental Organizations and Conferences

Environmental organizations can be global, regional, national or local; they can be government-run or private (NGO). Despite a tendency to see environmentalism as an American or Western-centered pursuit, almost every country has its share of environmental activism. Moreover, groups dedicated to community development and social justice may also attend to environmental concerns.

There are some volunteer organizations. For example Ecoworld, which is about the environment and is based in team work and volunteer work. The Wilderness Society, and global groups like the World Wide Fund for Nature and Friends of the Earth, disseminate information, participate in public hearings, lobby, stage demonstrations, and may purchase land for preservation. Smaller groups, including Wildlife Conservation International, conduct research on endangered species and ecosystems. More radical organizations, such as Greenpeace, Earth First!, and the Earth Liberation Front, have more directly opposed actions they regard as environmentally harmful. While Greenpeace is devoted to nonviolent confrontation as a means of bearing witness to environmental wrongs and bringing issues into the public realm for debate, the underground Earth Liberation Front engages in the clandestine destruction of property, the release of caged or penned animals, and other criminal acts. Such tactics are regarded as unusual within the movement, however.

On an international level, concern for the environment was the subject of a UN conference in Stockholm in 1972, attended by 114 nations. Out of this meeting developed UNEP (United Nations Environment Programme) and the follow-up United Nations Conference on Environment and Development in 1992. Other international organizations in support of environmental policies development include the Commission for Environmental Cooperation (NAFTA), the European Environment Agency (EEA), and the Inter-governmental Panel on Climate Change (IPCC).

Energy Conservation

Energy conservation is the practice of decreasing the quantity of energy used. It may be achieved through efficient energy use, in which case energy use is decreased while achieving a similar outcome, or by reduced consumption of energy services. Energy conservation may result in increase of financial capital, environmental value, national security, personal security, and human comfort. Individuals and organizations that are direct consumers of energy may want to conserve energy in order to reduce energy costs and promote economic security. Industrial and commercial users may want to increase efficiency and thus maximize profit.

Electrical energy conservation is an important element of energy policy. Energy conservation reduces the energy consumption and energy demand per capita and thus offsets some of the growth in energy supply needed to keep up with population growth. This reduces the rise in energy costs, and can reduce the need for new power plants, and energy imports. The reduced energy demand can provide more flexibility in choosing the most preferred methods of energy production.

By reducing emissions, energy conservation is an important part of lessening climate change. Energy conservation facilitates the replacement of non-renewable resources with renewable energy. Energy conservation is often the most economical solution to energy shortages, and is a more environmentally benign alternative to increased energy production.

Energy Conservation by Country

India

Bureau of Energy Efficiency is an Indian governmental organization created in 2002 responsible for promoting energy efficiency and conservation.

New Zealand

In New Zealand the Energy Efficiency and Conservation Authority is responsible for promoting energy efficiency and conservation.

United Kingdom

Energy conservation in the United Kingdom has been receiving increased attention over recent years. Key factors behind this are the Government's commitment to reducing carbon emissions, the projected 'energy gap' in UK electricity generation, and the increasing reliance on imports to meet national energy needs. Domestic housing and road transport are currently the two biggest problem areas.

The UK Government has jointly funded the Energy Saving Trust to promote energy conservation at a consumer, business and community level since 1993.

United States

The United States is currently the largest single consumer of energy. The U.S. Department of Energy categorizes national energy use in four broad sectors: transportation, residential, commercial, and industrial.

Energy usage in transportation and residential sectors (about half of U.S. energy consumption) is largely controlled by individual domestic consumers. Commercial and industrial energy expenditures are determined by businesses entities and other facility managers. National energy policy has a significant effect on energy usage across all four sectors.

Issues with Energy Conservation

Critics and advocates of some forms of energy conservation make the following arguments:

- Standard economic theory suggests that technological improvements that increase energy efficiency will tend to increase, rather than reduce energy use. This is called the *Jevons Paradox* and it is said to occur in two ways. Firstly, increased energy efficiency makes the use of energy relatively cheaper, thus encouraging increased use. Secondly, increased energy efficiency leads to increased economic growth, which pulls up energy use in the whole economy. This does not imply that increased fuel efficiency is worthless. Increased fuel efficiency enables greater production and a higher quality of life.
- Some retailers argue that bright lighting stimulates purchasing. Health studies have demonstrated that headache, stress, blood pressure, fatigue and worker error all generally increase with the common over-illumination present in many workplace and retail settings. It has been shown that natural daylighting increases productivity levels of workers, while reducing energy consumption.
- The use of telecommuting by major corporations is a significant opportunity to conserve energy, as many people now work in service jobs that enable them to work from home instead of commuting to work each day.
- Electric motors consume more than 60% of all electrical energy generated and are responsible for the loss of 10

to 20% of all electricity converted into mechanical energy.

- Consumers are often poorly informed of the savings of energy efficient products. The research one must put into conserving energy often is too time consuming and costly when there are cheaper products and technology available using today's fossil fuels.
- Technology needs to be able to change behavioural patterns, it can do this by allowing energy users, business and residential, to see graphically the impact their energy use can have in their workplace or homes. Advance real-time energy metering is able to help "people" save energy by their actions.

WASTE MINIMIZATION

Waste minimization is the process and the policy of reducing the amount of waste produced by a person or a society. It is part of the wider aim of waste reduction which is often described as a component of the waste hierarchy.

In the waste hierarchy, the most effective policies and processes are at the top. Waste minimization is also strongly related to efforts to minimize resource and energy use. For the same commercial output, usually the fewer materials are used, the less waste is produced. Waste minimization usually requires knowledge of the production process, cradle-to-grave analysis (the tracking of materials from their extraction to their return to earth) and detailed knowledge of the composition of the waste.

The main sources of waste vary from country to country. In the UK, most waste comes from the manufacturing industry, agriculture, construction and demolition industries. Household waste constitutes a relatively small proportion of all waste. Reasons for the creation of waste sometimes include requirements in the supply chain. For example, a company handling a product may insist that it should be packaged using particular packing because it fits its packaging equipment.

In contrast to waste minimization, waste management traditionally focuses on processing waste after it is created, concentrating on re-use, recycling, composting and waste-to-energy conversion.

In Industries

In industries, using more efficient manufacturing processes and better materials will generally reduce the production of waste. The applications of waste minimization have led to the development of innovative and commercially successful replacement products. Waste minimization has proven benefits to industry and the wider environment:

Waste minimization often requires investment, which is usually compensated by the savings. However, waste reduction in one part of the production process may create waste production in another part.

There are government incentives for waste minimization, which focus on the environmental benefits of adopting waste minimization strategies. In the UK, several pilot schemes such as The Catalyst Project and the Dee Waste Minimization Project, have shown the efficacy of such policies. Fourteen companies in Merseyside took part in the Catalyst Project; the project generated overall savings of £9 million and landfill waste was reduced by 12,000 tonnes per year.

The following is a list of waste minimization processes:

Resource Optimization

Minimizing the amount of waste produced by organizations or individuals goes hand-in-hand with optimizing their use of raw materials. For example, a dressmaker may arrange pattern pieces on a length of fabric in a particular way to enable the garment to be cut out from the smallest area of fabric.

Reuse of Scrap Material

The introduction of techniques or processes that enable production scrap to be immediately re-incorporated at the beginning of the manufacturing line so that they do not become a waste product. Many industries routinely do this; for example, paper mills return any damaged rolls to the beginning of the production line, and in the manufacture of plastic items, off-cuts and scrap are re-incorporated into new products.

Improved Quality Control and Process Monitoring

Taking steps to ensure that the number of reject batches is kept to a minimum. This is achieved by increasing the frequency of inspection and the number of points of inspection. For example, installing automated continuous monitoring equipment can help to identify production problems at an early stage.

Waste Exchanges

Where the waste product of one process becomes the raw material for a second process. Waste exchanges represent another way of reducing waste disposal volumes for waste that cannot be eliminated.

Ship to Point of Use

Making deliveries of incoming raw materials or components direct to the point where they are assembled or used in the manufacturing process can minimize handling and the use of protective wrappings or enclosures.

Product Design

Waste minimization and resource maximization for manufactured products can most easily be done at the design stage. Reducing the number of components used in a product or making the product easier to take apart can make it easier to be repaired or recycled at the end of its useful life.

In some cases, it may be best not to minimize the volume of raw materials used to make a product, but instead reduce the volume or toxicity of the waste created at the end of a product's life, or the environmental impact of the product's use.

Fitting the Intended Use

A product manufactured for "one off use" should be designed to meet its intended use. This applies especially to packaging materials, which should only be as durable as necessary to serve their intended purpose. In terms of food packaging, it would be wasteful if food, which has consumed resources and energy in its production, is damaged and spoiled due to extreme measures to reduce the use of paper, metals, glass and plastics in its packaging.

Durability

Improving product durability, such as extending a vacuum cleaner's useful life to 15 years instead of 12, can reduce waste and usually much improves resource optimization.

But in some cases it has a negative environmental impact. If a product is too durable, its replacement with more efficient technology is likely to be delayed. For example, a washing machine produced 10 years ago may use twice as much water, detergent and energy as one produced today. Therefore, extending an older machine's useful life may place a heavier burden on the environment than scrapping it, recycling its metal and buying a new model. Similarly, older vehicles consume more fuel and produce more emissions than their modern counterparts.

Some proponents of waste minimization consider that the way forward may be to view any manufactured product at the end of its useful life as a resource for recycling and reuse rather than waste. Recycling a product is easier if it is constructed of fewer materials. Car manufacturers have recently reduced the number of plastics used in their cars from twenty or more to three or four, hence simplifying the recovery of plastics from scrapped cars. However, exceptions (like having a combination of paper and plastic or plastic coating on glass) do exist, and might enable a product to fulfil its role with the minimum of resources.

Probably there will always be conflict between minimizing the resources used to make a product and the possibility of reusing or recycling it. For example, to make refillable glass bottles strong enough to withstand several journeys between the consumer and the bottling plant requires making them thicker and so heavier, which increases the resources required to transport them. Since transport has a large environmental impact, careful evaluation is required of the number of return journeys bottles make. If a refillable bottle is thrown away after being refilled only several times, the resources wasted may be greater than if the bottle had been designed for a single journey.

Many choices involve trade-offs of environmental impact, and often there is insufficient information to make informed decisions.

In Households

Household waste includes the refuse put out for collection, and the resources used in the home. Waste minimization may be complex for individuals and households, but there are some starting points that can be incorporated into the general household routine.

For example, appropriate amounts and sizes should be chosen when purchasing goods; buying large containers of paint for a small decorating job or buying larger amounts of food that must be consumed while still fresh create unnecessary waste. Also, if a pack or can is to be thrown away, food/liquid remains from the cans, packaging, ... must be removed

In a domestic situation, the potential for minimization is often dictated by lifestyle. Some people may view it as wasteful to purchase new products solely to follow fashion trends when the older products are still usable. Adults working full-time have little free time, and so may have to purchase more convenience foods that require little preparation, or prefer disposable nappies if there is a baby in the family.

Home composting, the practice of turning kitchen and garden waste into compost can be considered waste minimization. But other procedures that require the handling of waste, such as waste collected for recycling, while reducing the pressure on disposal facilities, do not reduce the amount of waste generated.

The resources that households use can be reduced considerably by using electricity thoughtfully (e.g. turning off lights and equipment when it is not needed) and by reducing the number of car journeys made. Individuals can reduce the amount of waste they create by buying fewer products and by buying products which last longer. Mending broken or worn items of clothing or equipment also contributes to minimizing household waste. Individuals may wish to consider minimizing their water usage, for example, or walk or cycle to their destination rather than using their car to save fuel and cut down emissions.

It can be wasteful to drive several miles to deposit a few newspapers, empty cans, glass, or plastic bottles into collection banks. The resources consumed, plus those needed to take the materials for reprocessing, could exceed the resources saved by the householder discarding the items in the general refuse collection.

The amount of waste an individual produces is a small portion of all waste produced by society, and personal waste reduction can only make a small impact on overall waste volumes. Yet, influence on policy can be exerted in other areas. Increased consumer awareness of the impact and power of certain purchasing decisions allows industry and individuals to change the total resource consumption. Consumers can influence manufacturers and distributors by avoiding buying products that do not have eco-labelling, which are currently not mandatory, or choosing products that minimize the use of packaging. Where reuse schemes are available, consumers can be proactive and use them.

Pull Apart

A packaging recycling classification system, that is based on the balance between a product's local kerbside recyclable and non-recyclable components. This ratio or difference is used as the basis for a grading system (PAC, abbreviation of PullApart Code), a survey and pressure concept, it was invented by Michael Butler of Dawlish, England, in 2005, and is operated for free. This system considers the environment and consumers to be of uppermost importance and seeks to modify packaging and recycling to serve these two masters. This disassembly of packaging, is also used locally, by the public, to freely determine and confirm which parts should go where, for their local authority installed, recycling bin refuse collection systems. According to PullApart's current Teignbridge (2009), survey of over 2100 products (constantly updated), just 3% are ideally suited for kerbside recycling, and broadly speaking a further 30% are good, with the rest failing.

POLLUTION PREVENTION

Pollution prevention (P2) describes activities that reduce the amount of pollution generated by a process, whether it is consumer consumption, driving, or industrial production. In contrast to most pollution control strategies, which seek to manage a pollutant after it is formed and reduce its impact upon the environment, the pollution prevention approach seeks to increase the efficiency of a process, thereby reducing the amount of pollution generated at its source. Although there is wide agreement that source reduction is the preferred strategy, some professionals also use the term pollution prevention to include recycling or reuse.

As an environmental management strategy, pollution prevention shares many attributes with cleaner production, a term used more commonly in other countries. Pollution prevention encompasses more specialized sub-disciplines including green chemistry and green design (also known as environmentally conscious design).

Reuse

Reuse is to use an item more than once. This includes conventional reuse where the item is used again for the same function, and new-life reuse where it is used for a new function. In contrast, recycling is the breaking down of the used item into raw materials which are used to make new items. By taking useful products and exchanging them, without reprocessing, reuse help us save time, money, energy and resources. In broader economic terrns, reuse offers quality products to people and organizations with limited means, while generating jobs and business activity that contribute to the economy.

Historically, financial motivation was one of the main drivers of reuse. In the developing world this driver can lead lead to very high levels of reuse, however rising wages and consequent consumer demand for the convenience of disposable products has made the reuse of low value items such as packaging uneconomic in richer countries, leading to the demise of many reuse programs. Current environmental awareness is gradually changing attitudes and regulations, such as the new packaging regulations, are gradually beginning to reverse the situation.

The classic example of conventional reuse is the doorstep delivery of milk in refillable bottles; other examples include the retreading of tires and the use of reusable plastic boxes (transit packing) instead of single-use cardboard cartons.

Advantages and Disadvantages

Reuse has certain potential advantages which can be summarized:

- Energy and raw materials savings as replacing many single use products with one reusable one reduces the number that need to be manufactured.
- Reduced disposal needs and costs.
- Refurbishment can bring sophisticated, sustainable, well paid jobs to underdeveloped economies.
- Cost savings for business and consumers as a reusable product is often cheaper than the many single use products it replaces.
- Some older items were better handcrafted and appreciate in value.

Disadvantages are also apparent:

- Reuse often requires cleaning or transport, which have environmental costs.
- Some items, such as freon appliances or infant auto seats, could be hazardous or less energy efficient as they continue to be used.
- Reusable products need to be more durable than single use products, and hence require more material per item. This is particularly significant if only a small proportion of the reusable products are in fact reused.

- Sorting and preparing items for reuse takes time, which is inconvenient for consumers and costs money for businesses.

Examples of Reuse

Reuse Centers and Virtual Exchanges

These services facilitate the transaction and redistribution of unwanted, yet perfectly usable, materials and equipment from one entity to another. The entities that benefit from either side of this service (as donors, sellers, recipients, or buyers) can be businesses, nonprofits, schools, community groups, and individuals. Some maintain a physical space (a reuse center), and others act as a matching service (a virtual exchange). Reuse centers generally maintain both warehouses and trucks. They take possession of the donated materials and make them available for redistribution or sale. Virtual exchanges do not have physical space or trucks, but instead allow users to post listings of materials available and wanted (for free or at low cost) on an online materials exchange website. Staff will help facilitate the exchange of these materials without ever taking possession of the materials.

In Australia Reverse Garbage in Sydney is the largest reuse centre in the country diverting more than 12,000 cubic metres of resources from landfill each year.

Remanufacturing

The cheapest reuse economies are "repair and overhaul" industries which take valuable parts, such as engine blocks, toner cartridges, "one use" cameras, aircraft hulls, and cathode ray tubes (CRTs) and refurbish them in a factory environment, hoping to meet the same specifications as new products. Xerox (copy machines), Video Display Corp.(CRTs) and Cummins Engine are examples of refurbishing factories in United States. Rolls Royce has a very large aircraft remanufacturing factory in Singapore; Caterpillar recently announced the opening of a tractor refurbishing plant in China. Some factories operate in competition with the original equipment manufacturer (OEM). When the refurbished item is resold under a new label (used monitor CRTs made into TVs, or cameras resold under a new label) this has been found legal by most courts.

When the item is resold under the same OEM name, it is informally considered a "gray market" item - if it is sold as used, it's legal, if it's represented as an OEM product eligible for rebates and warrantees, it is considered "counterfeit" or "black market". The automobile parts industry in the USA is governed by laws on the disclosure of "used" parts, and in some state mattresses which have been used are required to be sanitized or destroyed. Whether these laws are in place to protect consumers from black market items, or to protect manufacturers.

Deposit Programmes

These offer customers a financial incentive to return packaging for reuse. Although no longer common, international experience is showing that they can still be an effective way to encourage packaging reuse. However, financial incentive, unless great, may be less of an incentive than convenience: statistics show that, on average, a milk bottle is returned 12 times, whereas a lemonade bottle with a 15p deposit is returned, on average, only 3 times.

Refillable bottles are used extensively in many European countries; for example in Denmark, 98% of bottles are refillable, and 98% of those are returned by consumers. These systems are typically supported by deposit laws and other regulations.

Sainsbury Ltd have operated a plastic carrier bag cash refund scheme in 1991 - "the penny back scheme". The scheme is reported to save 970 tonnes of plastic per annum. The scheme has now been extended to a penny back on a voucher which can be contributed to schools registered on the scheme; it estimates this will raise the savings in plastic to 2500 tonnes per annum.

In some developing nations like India and Pakistan, the cost of new bottles often forces manufacturers to collect and refill old glass bottles for selling cola and other drinks. India and Pakistan also have a way of reusing old newspapers: "Kabadiwalas" buy these from the readers for scrap value and reuse them as packaging or recycle them. Scrap intermediaries help consumer dispose other other materials including metals and plastics.

Closed-Loop Programmes

These apply primarily to items of packaging, for example, where a company is involved in the regular transportation of goods from a central manufacturing facility to warehouses or warehouses to retail outlets. In these cases there is considerable benefit to using reusable "transport packaging" such as plastic crates or pallets. Tesco have established a series of nine recycling service units which wash returnable plastic trays; it is estimated that this operations saves around 50,000 tonnes of packaging per annum. Marks and Spencer operate a similar scheme with 90% reuse or recycle of transit packaging. 65% of their foods are transported on reusable plastic trays saving 25,000 tonnes of cardboard per year; they also have a 3 year plan to eliminate transit packaging on textiles and home furnishing product lines saving another 28,000 tonnes per annum. The same company started a coat hanger reuse scheme in 1993 and now reuse over 20 million of these annually saving 1,200 tonnes of plastic.

The benefits of closed loop reuse are primarily due to virtually no additional transport costs being involved, the empty lorry returning with the empty crates. There have been some recent attempts to get the public to join in on closed loop reuse schemes with the so called "blue basket" schemes (green in the case of Safeway) where shoppers use reusable plastic baskets in place of carrier bags for transporting their goods home from the supermarket; these baskets fit on specially designed trolleys making shopping supposedly easier.

Refilling Programmes

There have been some market led initiatives to encourage packaging reuse by companies introducing refill packs of certain commodities (mainly soap powders and cleaning fluids), the contents being transferred before use into a reusable package kept by the customer, with the savings in packaging being passed onto the customer by lower shelf prices. The refill pack itself is not reused, but being a minimal package for carrying the product home, it requires less material than one with the durability and features (reclosable top, convenient shape, etc) required for easy use of the product, while avoiding the transport cost and emissions of returning the reusable package to the factory.

Regiving/Regifting

Some items, such as clothes and children's toys, often become unwanted before they wear out due to changes in their owner's needs or preferences; these can be reused by selling or giving them to new owners. Regiving can take place informally between family, friends, or neighbours, through explicitly environmental organizations such as Freecycle and Freesharing Networks or through anti-poverty charities such as the Red Cross, United Way, Salvation Army, and Goodwill which give these items to those who could not afford them new. This has partly motivated movements such as The Compact, whose members promise not to buy anything new for a year, and rely on reusing items that otherwise would be thrown away. Reuse not only reduces Landfill in line with the Waste minimization programme but can help raise money for a good cause.

Printer Cartridges and Toners Reuse

Printer ink cartridges can be reused. They are sorted into different brands and models which may then be refilled, or resold back to the companies that created these cartridges. The companies then refill the ink reservoir which can be sold back to consumers. Toner cartridges are recycled the same way as ink cartridges, using toner instead of ink. This method is highly efficient as there is no energy spent on melting and recreating the cartridges.

Repurposing

Reuse is not limited to repeated uses for the same purpose. Examples of repurposing include using tires as boat fenders and steel drums as feeding troughs. Incinerator and power plant exhaust stack fly-ash is used extensively as an additive to concrete, providing increased strength. This type of reuse can sometimes make use of items which are no longer usable for their original purpose, for example using worn-out clothes as rags.

Waste Exchanges

A Waste Exchange, or Virtual Exchange facilitates the use of a waste product from one process as a raw material for another. As with new life reuse of finished items, this avoids the environmental costs of disposing of the waste and obtaining new raw material, and may still be possible if the nature of the process makes avoiding production of the waste or recycling it back into the original process impossible.

This sort of scheme needs to have a far broader base than is currently the case, it requires organization and the setting up of waste brokerages where lists of currently available wastes are and the quantities available. One of the problems is once a demand for a waste is known or shown then the material is no longer a "waste" but a sellable commodity which often prices itself out of the market, c.f waste cement kiln dust and N-viro (lime conditioned sewage sludge fertilizer). In the former East Germany, organic household waste was collected and used as fodder for pigs. This integrated system was made possible by the state's control of agriculture; the complexities of continuing it in a market economy after German reunification meant the system had to be discontinued.

Measuring the Impact of Reuse / Reuse Metrics

There are many ways of measuring the positive environmental, economic and social impact reuse has on our communities.

These include, but are not limited, to: - # of tons diverted from the landfill

- avoided dispoal costs (donor/seller)
- avoided purchase costs (recipient/buyer)
- value of materials donated (donor)
- revenues earned (donor/seller)
- of job created or retained
- of families/individuals/organizations assisted

Internalized Environmental Costs

This is an economist's way of saying introduce an environmental tax: a charge on items which reflects the environmental costs of their manufacture and disposal. This makes the environmental benefit of using one reusable item instead of many disposable ones into a financial incentive. Such charges have been introduced in some countries. Such schemes are said to encourage reuse.

Comparison to Recycling

Recycling differs from reuse in that it breaks down the item into raw materials which are then used to make new items, as opposed to reusing the intact item. As this extra processing requires energy, as a rule of thumb reuse is environmentally preferable to recycling ("reduce, reuse, recycle"), though recycling does have a significant part to play as it can often make use of items which are broken, worn out or otherwise unsuitable for reuse. However, as transport emissions are significant portion of the environmental impact of both reuse and recycling, in some cases recycling is the more prudent course as reuse can require long transport distances. A complex lifecycle analysis may be required during a products design phase to determine the efficacy of reuse, recycling, or neither, and produce accordingly.

Reuse of Information

Besides the great amount of reuse of our physical resources, there is now a powerful argument for reuse of information, notably program code for the software that drives our computers and the Internet, but also the documentation that explains how to use every modern device. And it is proposed as a way to improve education by assembling a great library of shareable learning objects that can be reused in learning management systems.

Software reuse grew out of the standard subroutine libraries of the 1960s. It is the main principle of today's object-oriented programming. Instead of constantly reinventing software wheels, programming languages like C++, Java, Objective C, and others are building vast collections of reusable software objects and components.

11
Ozone Depletion and Global Warming

Ozone depletion describes two distinct, but related observations: a slow, steady decline of about 4% per decade in the total volume of ozone in Earth's stratosphere (ozone layer) since the late 1970s, and a much larger, but seasonal, decrease in stratospheric ozone over Earth's polar regions during the same period. The latter phenomenon is commonly referred to as the ozone hole. In addition to this well-known stratospheric ozone depletion, there are also tropospheric ozone depletion events, which occur near the surface in polar regions during spring.

The detailed mechanism by which the polar ozone holes form is different from that for the mid-latitude thinning, but the most important process in both trends is catalytic destruction of ozone by atomic chlorine and bromine. The main source of these halogen atoms in the stratosphere is photodissociation of chlorofluorocarbon (CFC) compounds, commonly called freons, and of bromofluorocarbon compounds known as halons. These compounds are transported into the stratosphere after being emitted at the surface. Both ozone depletion mechanisms strengthened as emissions of CFCs and halons increased.

CFCs and other contributory substances are commonly referred to as ozone-depleting substances (ODS). Since the ozone layer prevents most harmful UVB wavelengths (270–315 nm) of ultraviolet light (UV light) from passing through the Earth's atmosphere, observed and projected decreases in ozone have generated worldwide concern leading to adoption of the Montreal Protocol banning the production of CFCs and halons as well as related ozone depleting chemicals such as carbon tetrachloride and trichloroethane. It is suspected that a variety of biological consequences such as increases in skin cancer, damage to plants, and reduction of plankton populations in the ocean's photic zone may result from the increased UV exposure due to ozone depletion.

Ozone Cycle Overview

Three forms (or allotropes) of oxygen are involved in the ozone-oxygen cycle: oxygen atoms (O or atomic oxygen), oxygen gas (O_2 or diatomic oxygen), and ozone gas (O_3 or triatomic oxygen). Ozone is formed in the stratosphere when oxygen molecules photodissociate after absorbing an ultraviolet photon whose wavelength is shorter than 240 nm. This produces two oxygen atoms. The atomic oxygen then combines with O_2 to create O_3. Ozone molecules absorb UV light between 310 and 200 nm, following which ozone splits into a molecule of O_2 and an oxygen atom. The oxygen atom then joins up with an oxygen molecule to regenerate ozone. This is a continuing process which terminates when an oxygen atom "recombines" with an ozone molecule to make two O_2 molecules:

$$O + O_3 \rightarrow 2\ O_2$$

Ozone can be destroyed by a number of free radical catalysts, the most important of which are the hydroxyl radical (OH.), the nitric oxide radical (NO.) and atomic chlorine (Cl.) and bromine (Br.). All of these have both natural and anthropogenic (manmade) sources; at the present time, most of the OH. and NO. in the stratosphere is of natural origin, but human activity has dramatically increased the levels of chlorine and bromine. These elements are found in certain stable organic compounds, especially chlorofluorocarbons

(CFCs), which may find their way to the stratosphere without being destroyed in the troposphere due to their low reactivity. Once in the stratosphere, the Cl and Br atoms are liberated from the parent compounds by the action of ultraviolet light, e.g. ('h' is Planck's constant, 'í' is frequency of electromagnetic radiation)

$$CFCl_3 + hí \rightarrow CFCl_2 + Cl$$

The Cl and Br atoms can then destroy ozone molecules through a variety of catalytic cycles. In the simplest example of such a cycle, a chlorine atom reacts with an ozone molecule, taking an oxygen atom with it (forming ClO) and leaving a normal oxygen molecule. The chlorine monoxide (i.e., the ClO) can react with a second molecule of ozone (i.e., O_3) to yield another chlorine atom and two molecules of oxygen. The chemical shorthand for these gas-phase reactions is:

$$Cl + O_3 \rightarrow ClO + O_2$$

$$ClO + O_3 \rightarrow Cl + 2\ O_2$$

The overall effect is a decrease in the amount of ozone. More complicated mechanisms have been discovered that lead to ozone destruction in the lower stratosphere as well.

A single chlorine atom would keep on destroying ozone (thus a catalyst) for up to two years (the time scale for transport back down to the troposphere) were it not for reactions that remove them from this cycle by forming reservoir species such as hydrogen chloride (HCl) and chlorine nitrate ($ClONO_2$). On a per atom basis, bromine is even more efficient than chlorine at destroying ozone, but there is much less bromine in the atmosphere at present. As a result, both chlorine and bromine contribute significantly to the overall ozone depletion. Laboratory studies have shown that fluorine and iodine atoms participate in analogous catalytic cycles. However, in the Earth's stratosphere, fluorine atoms react rapidly with water and methane to form strongly-bound HF, while organic molecules which contain iodine react so rapidly in the lower atmosphere that they do not reach the stratosphere in significant quantities. Furthermore, a single chlorine atom is able to react with 100,000 ozone molecules. This fact plus the amount of chlorine released into the atmosphere by chlorofluorocarbons (CFCs) yearly demonstrates how dangerous CFCs are to the environment.

Quantitative Understanding of the Chemical Ozone Loss Process

New research on the breakdown of a key molecule in these ozone-depleting chemicals, dichlorine peroxide (Cl_2O_2), calls into question the completeness of present atmospheric models of polar ozone depletion. Specifically, chemists at NASA's Jet Propulsion Laboratory in Pasadena, California, found in 2007 that the temperatures, and the spectrum and intensity of radiation present in the stratosphere created conditions insufficient to allow the rate of chemical-breakdown required to release chlorine radicals in the volume necessary to explain observed rates of ozone depletion. Instead, laboratory tests, designed to be the most accurate reflection of stratospheric conditions to date, showed the decay of the crucial molecule almost a magnitude lower than previously thought.

That result motivated further measurements by different methods, resulting in cross-sections that agree with the older, higher ones. If the new results hold up the purported discrepancy will be resolved.

Observations on Ozone Layer Depletion

The most pronounced decrease in ozone has been in the lower stratosphere. However, the ozone hole is most usually measured not in terms of ozone concentrations at these levels (which are typically of a few parts per million) but by reduction in the total *column ozone*, above a point on the Earth's surface, which is normally expressed in Dobson units, abbreviated as "DU". Marked decreases in column ozone in the Antarctic spring and early summer compared to the early 1970s and before have been observed using instruments such as the Total Ozone Mapping Spectrometer (TOMS).

Reductions of up to 70% in the ozone column observed in the austral (southern hemispheric) spring over Antarctica and first reported in 1985 (Farman et al. 1985) are continuing. Through the 1990s, total column ozone in September and October have continued to be 40–50% lower than pre-ozone-hole values. In the Arctic the amount lost is more variable year-to-year than in the Antarctic. The greatest declines, up to 30%, are in the winter and spring, when the stratosphere is colder.

Reactions that take place on polar stratospheric clouds (PSCs) play an important role in enhancing ozone depletion. PSCs form more readily in the extreme cold of Antarctic stratosphere. This is why ozone holes first formed, and are deeper, over Antarctica. Early models failed to take PSCs into account and predicted a gradual global depletion, which is why the sudden Antarctic ozone hole was such a surprise to many scientists.

In middle latitudes it is preferable to speak of ozone depletion rather than holes. Declines are about 3% below pre-1980 values for 35–60°N and about 6% for 35–60°S. In the tropics, there are no significant trends.

Ozone depletion also explains much of the observed reduction in stratospheric and upper tropospheric temperatures. The source of the warmth of the stratosphere is the absorption of UV radiation by ozone, hence reduced ozone leads to cooling. Some stratospheric cooling is also predicted from increases in greenhouse gases such as CO_2; however the ozone-induced cooling appears to be dominant.

Predictions of ozone levels remain difficult. The World Meteorological Organization Global Ozone Research and Monitoring Project - Report No. 44 comes out strongly in favour for the Montreal Protocol, but notes that a UNEP 1994 Assessment overestimated ozone loss for the 1994–1997 period.

Chemicals in the Atmosphere

CFCs in the Atmosphere

Chlorofluorocarbons (CFCs) were invented by Thomas Midgley in the 1920s. They were used in air conditioning/ cooling units, as aerosol spray propellants prior to the 1980s, and in the cleaning processes of delicate electronic equipment. They also occur as by-products of some chemical processes. No significant natural sources have ever been identified for these compounds — their presence in the atmosphere is due almost entirely to human manufacture. As mentioned in the *ozone cycle overview* above, when such ozone-depleting chemicals reach the stratosphere, they are dissociated by ultraviolet light to release chlorine atoms. The chlorine atoms act as a catalyst, and each can break-down tens of thousands of ozone molecules before being removed from the stratosphere. Given the longevity of CFC molecules, recovery times are measured in decades. It is calculated that a CFC molecule takes an average of 15 years to go from the ground level up to the upper atmosphere, and it can stay there for about a century, destroying up to one hundred thousand ozone molecules during that time.

Verification of Observations

Scientists have been increasingly able to attribute the observed ozone depletion to the increase of anthropogenic halogen compounds from CFCs by the use of complex chemistry transport models and their validation against observational data (e.g. SLIMCAT, CLaMS). These models work by combining satellite measurements of chemical concentrations and meteorological fields with chemical reaction rate constants obtained in lab experiments. They are able to identify not only the key chemical reactions but also the transport processes which bring CFC photolysis products into contact with ozone.

The Ozone Hole and its Causes

The Antarctic ozone hole is an area of the Antarctic stratosphere in which the recent ozone levels have dropped to as low as 33% of their pre-1975 values. The ozone hole occurs during the Antarctic spring, from September to early December, as strong westerly winds start to circulate around the continent and create an atmospheric container. Within this polar vortex, over 50% of the lower stratospheric ozone is destroyed during the Antarctic spring.

As explained above, the overall cause of ozone depletion is the presence of chlorine-containing source gases (primarily CFCs and related halocarbons). In the presence of UV light, these gases dissociate, releasing chlorine atoms, which then go on to catalyze ozone destruction. The Cl-catalyzed ozone depletion can take place in the gas phase, but it is dramatically enhanced in the presence of polar stratospheric clouds (PSCs).

These polar stratospheric clouds form during winter, in the extreme cold. Polar winters are dark, consisting of 3 months without solar radiation (sunlight). Not only lack of sunlight contributes to a decrease in temperature but also the polar vortex traps and chills air. Temperatures hover around or below -80 °C. These low temperatures form cloud particles and are composed of either nitric acid (Type I PSC) or ice (Type II PSC). Both types provide surfaces for chemical reactions that lead to ozone destruction.

The photochemical processes involved are complex but well understood. The key observation is that, ordinarily, most of the chlorine in the stratosphere resides in stable "reservoir" compounds, primarily hydrochloric acid (HCl) and chlorine nitrate ($ClONO_2$). During the Antarctic winter and spring, however, reactions on the surface of the polar stratospheric cloud particles convert these "reservoir" compounds into reactive free radicals (Cl and ClO). The clouds can also remove NO_2 from the atmosphere by converting it to nitric acid, which prevents the newly formed ClO from being converted back into $ClONO_2$.

The role of sunlight in ozone depletion is the reason why the Antarctic ozone depletion is greatest during spring. During winter, even though PSCs are at their most abundant, there is no light over the pole to drive the chemical reactions. During the spring, however, the sun comes out, providing energy to drive photochemical reactions, and melt the polar stratospheric clouds, releasing the trapped compounds.

Most of the ozone that is destroyed is in the lower stratosphere, in contrast to the much smaller ozone depletion through homogeneous gas phase reactions, which occurs primarily in the upper stratosphere.

Warming temperatures near the end of spring break up the vortex around mid-December. As warm, ozone-rich air flows in from lower latitudes, the PSCs are destroyed, the ozone depletion process shuts down, and the ozone hole closes.

Interest in Ozone Layer Depletion

While the effect of the Antarctic ozone hole in decreasing the global ozone is relatively small, estimated at about 4% per decade, the hole has generated a great deal of interest because:

- The decrease in the ozone layer was predicted in the early 1980s to be roughly 7% over a 60 year period.
- The sudden recognition in 1985 that there was a substantial "hole" was widely reported in the press.

The especially rapid ozone depletion in Antarctica had previously been dismissed as a measurement error.

- Many were worried that ozone holes might start to appear over other areas of the globe but to date the only other large-scale depletion is a smaller ozone "dimple" observed during the Arctic spring over the North Pole. Ozone at middle latitudes has declined, but by a much smaller extent (about 4–5% decrease).
- If the conditions became more severe (cooler stratospheric temperatures, more stratospheric clouds, more active chlorine), then global ozone may decrease at a much greater pace. Standard global warming theory predicts that the stratosphere will cool.
- When the Antarctic ozone hole breaks up, the ozone-depleted air drifts out into nearby areas. Decreases in the ozone level of up to 10% have been reported in New Zealand in the month following the break-up of the Antarctic ozone hole.

Consequences of Ozone Layer Depletion

Since the ozone layer absorbs UVB ultraviolet light from the Sun, ozone layer depletion is expected to increase surface UVB levels, which could lead to damage, including increases in skin cancer. This was the reason for the Montreal Protocol. Although decreases in stratospheric ozone are well-tied to CFCs and there are good theoretical reasons to believe that decreases in ozone will lead to increases in surface UVB, there is no direct observational evidence linking ozone depletion to higher incidence of skin cancer in human beings. This is partly due to the fact that UVA, which has also been implicated in some forms of skin cancer, is not absorbed by ozone, and it is nearly impossible to control statistics for lifestyle changes in the populace.

Increased UV

Ozone, while a minority constituent in the Earth's atmosphere, is responsible for most of the absorption of UVB radiation. The amount of UVB radiation that penetrates through the ozone layer decreases exponentially with the slant-path thickness/density of the layer. Correspondingly, a decrease in atmospheric ozone is expected to give rise to significantly increased levels of UVB near the surface.

Increases in surface UVB due to the ozone hole can be partially inferred by radiative transfer model calculations, but cannot be calculated from direct measurements because of the lack of reliable historical (pre-ozone-hole) surface UV data, although more recent surface UV observation measurement programmes exist (e.g. at Lauder, New Zealand).

Because it is this same UV radiation that creates ozone in the ozone layer from O_2 (regular oxygen) in the first place, a reduction in stratospheric ozone would actually tend to increase photochemical production of ozone at lower levels (in the troposphere), although the overall observed trends in total column ozone still show a decrease, largely because ozone produced lower down has a naturally shorter photochemical lifetime, so it is destroyed before the concentrations could reach a level which would compensate for the ozone reduction higher up.

Biological Effects of Increased UV and Microwave Radiation from a Depleted Ozone Layer

The main public concern regarding the ozone hole has been the effects of surface UV on human health. So far, ozone depletion in most locations has been typically a few percent and, as noted above, no direct evidence of health damage is available in most latitudes. Were the high levels of depletion seen in the ozone hole ever to be common across the globe, the effects could be substantially more dramatic. As the ozone hole over Antarctica has in some instances grown so large as to reach southern parts of Australia and New Zealand, environmentalists have been concerned that the increase in surface UV could be significant.

Effects of Ozone Layer Depletion on Humans

UVB (the higher energy UV radiation absorbed by ozone) is generally accepted to be a contributory factor to skin cancer. In addition, increased surface UV leads to increased tropospheric ozone, which is a health risk to humans. The increased surface UV also represents an increase in the vitamin D synthetic capacity of the sunlight.

The cancer preventive effects of vitamin D represent a possible beneficial effect of ozone depletion. In terms of health costs, the possible benefits of increased UV irradiance may outweigh the burden.

1. *Basal and Squamous Cell Carcinomas* — The most common forms of skin cancer in humans, basal and squamous cell carcinomas, have been strongly linked to UVB exposure. The mechanism by which UVB induces these cancers is well understood — absorption of UVB radiation causes the pyrimidine bases in the DNA molecule to form dimers, resulting in transcription errors when the DNA replicates. These cancers are relatively mild and rarely fatal, although the treatment of squamous cell carcinoma sometimes requires extensive reconstructive surgery. By combining epidemiological data with results of animal studies, scientists have estimated that a one percent decrease in stratospheric ozone would increase the incidence of these cancers by 2%.

2. *Malignant Melanoma* — Another form of skin cancer, malignant melanoma, is much less common but far more dangerous, being lethal in about 15% - 20% of

the cases diagnosed. The relationship between malignant melanoma and ultraviolet exposure is not yet well understood, but it appears that both UVB and UVA are involved. Experiments on fish suggest that 90 to 95% of malignant melanomas may be due to UVA and visible radiation whereas experiments on opossums suggest a larger role for UVB. Because of this uncertainty, it is difficult to estimate the impact of ozone depletion on melanoma incidence. One study showed that a 10% increase in UVB radiation was associated with a 19% increase in melanomas for men and 16% for women. A study of people in Punta Arenas, at the southern tip of Chile, showed a 56% increase in melanoma and a 46% increase in nonmelanoma skin cancer over a period of seven years, along with decreased ozone and increased UVB levels.

3. *Cortical Cataracts* – Studies are suggestive of an association between ocular cortical cataracts and UV-B exposure, using crude approximations of exposure and various cataract assessment techniques. A detailed assessment of ocular exposure to UV-B was carried out in a study on Chesapeake Bay Watermen, where increases in average annual ocular exposure were associated with increasing risk of cortical opacity . In this highly exposed group of predominantly white males, the evidence linking cortical opacities to sunlight exposure was the strongest to date. However, subsequent data from a population-based study in Beaver Dam, WI suggested the risk may be confined to men. In the Beaver Dam study, the exposures among women were lower than exposures among men, and no association was seen. Moreover, there were no data linking sunlight exposure to risk of cataract in African Americans, although other eye diseases have different prevalences among the different racial groups, and

cortical opacity appears to be higher in African Americans compared with whites.

4. *Increased Tropospheric Ozone* — Increased surface UV leads to increased tropospheric ozone. Ground-level ozone is generally recognized to be a health risk, as ozone is toxic due to its strong oxidant properties. At this time, ozone at ground level is produced mainly by the action of UV radiation on combustion gases from vehicle exhausts.

Effects on Crops

An increase of UV radiation would be expected to affect crops. A number of economically important species of plants, such as rice, depend on cyanobacteria residing on their roots for the retention of nitrogen. Cyanobacteria are sensitive to UV light and they would be affected by its increase.

Effects on Plankton

Research has shown a widespread extinction of plankton 2 million years ago that coincided with a nearby supernova. There is a difference in the orientation and motility of planktons when excess of UV rays reach earth. Researchers speculate that the extinction was caused by a significant weakening of the ozone layer at that time when the radiation from the supernova produced nitrogen oxides that catalyzed the destruction of ozone (plankton are particularly susceptible to effects of UV light, and are vitally important to marine food webs).

Public Policy in Response to the Ozone Hole

The full extent of the damage that CFCs have caused to the ozone layer is not known and will not be known for decades; however, marked decreases in column ozone have already been observed.

After a 1976 report by the U.S. National Academy of Sciences concluded that credible scientific evidence supported the ozone depletion hypothesis, a few countries, including the United States, Canada, Sweden, and Norway, moved to eliminate the use of CFCs in aerosol spray cans. At the time this was widely regarded as a first step towards a more comprehensive regulation policy, but progress in this direction slowed in subsequent years, due to a combination of political factors (continued resistance from the halocarbon industry and a general change in attitude towards environmental regulation during the first two years of the Reagan administration) and scientific developments (subsequent National Academy assessments which indicated that the first estimates of the magnitude of ozone depletion had been overly large). The European Community rejected proposals to ban CFCs in aerosol sprays while even in the U.S., CFCs continued to be used as refrigerants and for cleaning circuit boards. Worldwide CFC production fell sharply after the U.S. aerosol ban, but by 1986 had returned nearly to its 1976 level. In 1980, DuPont closed down its research program into halocarbon alternatives.

The US Government's attitude began to change again in 1983, when William Ruckelshaus replaced Anne M. Burford as Administrator of the United States Environmental Protection Agency. Under Ruckelshaus and his successor, Lee Thomas, the EPA pushed for an international approach to halocarbon regulations. In 1985 20 nations, including most of the major CFC producers, signed the Vienna Convention for the Protection of the Ozone Layer which established a framework for negotiating international regulations on ozone-depleting substances. That same year, the discovery of the Antarctic ozone hole was announced, causing a revival in public attention to the issue. In 1987, representatives from 43 nations signed the Montreal Protocol. Meanwhile, the halocarbon industry shifted its position and started supporting a protocol to limit CFC production. The reasons for this were in part explained by "Dr. Mostafa Tolba, former head of the UN Environment Programme, who was quoted in the 30 June 1990 edition of The New Scientist, '...the chemical industry supported the Montreal Protocol in 1987 because it set up a worldwide schedule for phasing out CFCs, which [were] no longer protected by patents. This provided companies with an equal opportunity to market new, more profitable compounds.'"

At Montreal, the participants agreed to freeze production of CFCs at 1986 levels and to reduce production by 50% by 1999. After a series of scientific expeditions to the Antarctic produced convincing evidence that the ozone hole was indeed caused by chlorine and bromine from manmade organohalogens, the Montreal Protocol was strengthened at a 1990 meeting in London. The participants agreed to phase out CFCs and halons entirely (aside from a very small amount marked for certain "essential" uses, such as asthma inhalers) by 2000. At a 1992 meeting in Copenhagen, the phase out date was moved up to 1996.

To some extent, CFCs have been replaced by the less damaging hydro-chloro-fluoro-carbons (HCFCs), although concerns remain regarding HCFCs also. In some applications, hydro-fluoro-carbons (HFCs) have been used to replace CFCs. HFCs, which contain no chlorine or bromine, do not contribute at all to ozone depletion although they are potent greenhouse gases. The best known of these compounds is probably HFC-134a (R-134a), which in the United States has largely replaced CFC-12 (R-12) in automobile air conditioners. In laboratory analytics (a former "essential" use) the ozone depleting substances can be replaced with various other solvents.

Ozone Diplomacy, by Richard Benedick (Harvard University Press, 1991) gives a detailed account of the negotiation process that led to the Montreal Protocol. Pielke and Betsill provide an extensive review of early US government responses to the emerging science of ozone depletion by CFCs.

Current Events and Future Prospects of Ozone Depletion

Since the adoption and strengthening of the Montreal Protocol has led to reductions in the emissions of CFCs, atmospheric concentrations of the most significant compounds have been declining. These substances are being gradually removed from the atmosphere. By 2015, the Antarctic ozone hole would have reduced by only 1 million km^2 out of 25 (Newman *et al.*, 2004); complete recovery of the Antarctic ozone layer will not occur until the year 2050 or later. Work has suggested that a detectable (and statistically significant) recovery will not occur until around 2024, with ozone levels recovering to 1980 levels by around 2068.

The decrease in ozone-depleting chemicals has also been significantly affected by a decrease in bromine-containing chemicals. The data suggest that substantial natural sources exist for atmospheric methyl bromide (CH_3Br).

The 2004 ozone hole ended in November 2004, daily minimum stratospheric temperatures in the Antarctic lower stratosphere increased to levels that are too warm for the formation of polar stratospheric clouds (PSCs) about 2 to 3 weeks earlier than in most recent years.

The Arctic winter of 2005 was extremely cold in the stratosphere; PSCs were abundant over many high-latitude areas until dissipated by a big warming event, which started in the upper stratosphere during February and spread throughout the Arctic stratosphere in March. The size of the Arctic area of anomalously low total ozone in 2004-2005 was larger than in any year since 1997. The predominance of anomalously low total ozone values in the Arctic region in the winter of 2004-2005 is attributed to the very low stratospheric temperatures and meteorological conditions favorable for ozone destruction along with the continued presence of ozone destroying chemicals in the stratosphere.

A 2005 IPCC summary of ozone issues observed that observations and model calculations suggest that the global average amount of ozone depletion has now approximately stabilized. Although considerable variability in ozone is expected from year to year, including in polar regions where depletion is largest, the ozone layer is expected to begin to recover in coming decades due to declining ozone-depleting substance concentrations, assuming full compliance with the Montreal Protocol.

Temperatures during the Arctic winter of 2006 stayed fairly close to the long-term average until late January, with minimum readings frequently cold enough to produce PSCs. During the last week of January, however, a major warming event sent temperatures well above normal—much too warm to support PSCs. By the time temperatures dropped back to near normal in March, the seasonal norm was well above the PSC threshold. Preliminary satellite instrument-generated ozone maps show seasonal ozone buildup slightly below the long-term means for the Northern Hemisphere as a whole, although some high ozone events have occurred. During March 2006, the Arctic stratosphere poleward of 60 degrees North Latitude was free of anomalously low ozone areas except during the three-day period from 17 March to 19 when the total ozone cover fell below 300 DU over part of the North Atlantic region from Greenland to Scandinavia.

The area where total column ozone is less than 220 DU (the accepted definition of the boundary of the ozone hole) was relatively small until around 20 August 2006. Since then the ozone hole area increased rapidly, peaking at 29 million km^2 24 September. In October 2006, NASA reported that the year's ozone hole set a new area record with a daily average of 26 million km² between 7 September and 13 October 2006; total ozone thicknesses fell as low as 85 DU on 8 October. The two factors combined, 2006 sees the worst level of depletion in recorded ozone history. The depletion is attributed to the temperatures above the Antarctic reaching the lowest recording since comprehensive records began in 1979.

On October 2008 the Ecuadorian Space Agency published a report called HIPERION, a study of the last 28 years data from 10 satellites and dozens of ground instruments around the world among them their own, and found that the UV radiation reaching equatorial latitudes was far greater than expected, climbing in some very populated cities up to 24 UVI, the WHO UV Index standard considers 11 as an extreme index and a great risk to health. The report concluded that the ozone depletion around mid latitudes on the planet is already endangering large populations in this areas. Later, the CONIDA, the Peruvian Space Agency, made its own study, which found almost the same facts as the Ecuadorian study.

The Antarctic ozone hole is expected to continue for decades. Ozone concentrations in the lower stratosphere over Antarctica will increase by 5%–10% by 2020 and return to pre-1980 levels by about 2060–2075, 10–25 years later than predicted in earlier assessments. This is because of revised estimates of atmospheric concentrations of Ozone Depleting Substances — and a larger predicted future usage in developing countries. Another factor which may aggravate ozone depletion is the draw-down of nitrogen oxides from above the stratosphere due to changing wind patterns.

History of the Research

The basic physical and chemical processes that lead to the formation of an ozone layer in the Earth's stratosphere were discovered by Sydney Chapman in 1930. These are discussed in the article Ozone-oxygen cycle — briefly, short-wavelength UV radiation splits an oxygen (O_2) molecule into two oxygen (O) atoms, which then combine with other oxygen molecules to form ozone. Ozone is removed when an oxygen atom and an ozone molecule "recombine" to form two oxygen molecules, i.e. $O + O_3$ '! $2O_2$. In the 1950s, David Bates and Marcel Nicolet presented evidence that various free radicals, in particular hydroxyl (OH) and nitric oxide (NO), could catalyze this recombination reaction, reducing the overall amount of ozone. These free radicals were known to be present in the stratosphere, and so were regarded as part of the natural balance – it was estimated that in their absence, the ozone layer would be about twice as thick as it currently is.

In 1970, Prof. Paul Crutzen pointed out that emissions of *nitrous* oxide (N_2O), a stable, long-lived gas produced by soil bacteria, from the Earth's surface could affect the amount of *nitric* oxide (NO) in the stratosphere. Crutzen showed that nitrous oxide lives long enough to reach the stratosphere, where it is converted into NO. Crutzen then noted that increasing use of fertilizers might have led to an increase in nitrous oxide emissions over the natural background, which would in turn result in an increase in the amount of NO in the stratosphere. Thus human activity could have an impact on the stratospheric ozone layer. In the following year, Crutzen and (independently) Harold Johnston suggested that NO emissions from supersonic aircraft, which fly in the lower stratosphere, could also deplete the ozone layer.

The Rowland-Molina Hypothesis

In 1974, Frank Sherwood Rowland, Chemistry Professor at the University of California at Irvine, and his postdoctoral associate Mario J. Molina suggested that long-lived organic halogen compounds, such as CFCs, might behave in a similar fashion as Crutzen had proposed for nitrous oxide. James Lovelock (most popularly known as the creator of the Gaia hypothesis) had recently discovered, during a cruise in the South Atlantic in 1971, that almost all of the CFC compounds manufactured since their invention in 1930 were still present in the atmosphere. Molina and Rowland concluded that, like N_2O, the CFCs would reach the stratosphere where they would be dissociated by UV light, releasing Cl atoms. (A year earlier, Richard Stolarski and Ralph Cicerone at the University of Michigan had shown that Cl is even more efficient than NO at catalyzing the destruction of ozone. Similar conclusions were reached by Michael McElroy and Steven Wofsy at Harvard University. Neither group, however, had realized that CFC's were a potentially large source of stratospheric chlorine — instead, they had been investigating the possible effects of HCl emissions from the Space Shuttle, which are very much smaller.)

The Rowland-Molina hypothesis was strongly disputed by representatives of the aerosol and halocarbon industries. The Chair of the Board of DuPont was quoted as saying that ozone depletion theory is "a science fiction tale...a load of rubbish...utter nonsense". Robert Abplanalp, the President of Precision Valve Corporation (and inventor of the first practical aerosol spray can valve), wrote to the Chancellor of UC Irvine to complain about Rowland's public statements. Nevertheless, within three years most of the basic assumptions made by Rowland and Molina were confirmed by laboratory measurements and by direct observation in the stratosphere. The concentrations of the source gases (CFCs and related compounds) and the chlorine reservoir species (HCl and $ClONO_2$) were measured throughout the stratosphere, and demonstrated that CFCs were indeed the major source of stratospheric chlorine, and that nearly all of the CFCs emitted would eventually reach the stratosphere. Even more convincing was the measurement, by James G. Anderson and collaborators, of chlorine monoxide (ClO) in the stratosphere. ClO is produced by the reaction of Cl with ozone — its observation thus demonstrated that Cl radicals not only were present in the stratosphere but also were actually involved in destroying ozone. McElroy and Wofsy extended the work of Rowland and Molina by showing that bromine atoms were even more effective catalysts for ozone loss than chlorine atoms and argued that the brominated organic compounds known as halons, widely used in fire extinguishers, were a potentially large source of stratospheric bromine. In 1976 the U.S. National Academy of Sciences released a report which concluded that the ozone depletion hypothesis was strongly supported by the scientific evidence. Scientists calculated that if CFC production continued to increase at the going rate of 10% per year until 1990 and then remain steady, CFCs would cause a global ozone loss of 5 to 7% by 1995, and a 30 to 50% loss by 2050. In response the United States, Canada and Norway banned the use of CFCs in aerosol spray cans in 1978. However, subsequent research, summarized by the National Academy in reports issued between 1979 and 1984, appeared to show that the earlier estimates of global ozone loss had been too large.

Crutzen, Molina, and Rowland were awarded the 1995 Nobel Prize in Chemistry for their work on stratospheric ozone.

The Ozone Hole

The discovery of the Antarctic "ozone hole" by British Antarctic Survey scientists Farman, Gardiner and Shanklin (announced in a paper in *Nature* in May 1985) came as a shock to the scientific community, because the observed decline in polar ozone was far larger than anyone had anticipated. Satellite measurements showing massive depletion of ozone around the south pole were becoming available at the same time. However, these were initially rejected as unreasonable by data quality control algorithms (they were filtered out as errors since the values were unexpectedly low); the ozone hole was detected only in satellite data when the raw data was reprocessed following evidence of ozone depletion in *in situ* observations. When the software was rerun without the flags, the ozone hole was seen as far back as 1976.

Susan Solomon, an atmospheric chemist at the National Oceanic and Atmospheric Administration (NOAA), proposed that chemical reactions on polar stratospheric clouds (PSCs) in the cold Antarctic stratosphere caused a massive, though localized and seasonal, increase in the amount of chlorine present in active, ozone-destroying forms. The polar stratospheric clouds in Antarctica are only formed when there are very low temperatures, as low as -80 degrees C, and early spring conditions. In such conditions the ice crystals of the cloud provide a suitable surface for conversion of unreactive chlorine compounds into reactive chlorine compounds which can deplete ozone easily.

Moreover, the polar vortex formed over Antarctica is very tight and the reaction which occurs on the surface of the cloud crystals is far different from when it occurs in atmosphere. These conditions have led to ozone hole formation in Antarctica. This hypothesis was decisively confirmed, first by laboratory measurements and subsequently by direct measurements, from the ground and from high-altitude airplanes, of very high concentrations of chlorine monoxide (ClO) in the Antarctic stratosphere.

Alternative hypotheses, which had attributed the ozone hole to variations in solar UV radiation or to changes in atmospheric circulation patterns, were also tested and shown to be untenable.

Meanwhile, analysis of ozone measurements from the worldwide network of ground-based Dobson spectrophotometers led an international panel to conclude that the ozone layer was in fact being depleted, at all latitudes outside of the tropics. These trends were confirmed by satellite measurements. As a consequence, the major halocarbon producing nations agreed to phase out production of CFCs, halons, and related compounds, a process that was completed in 1996.

Since 1981 the United Nations Environment Programme has sponsored a series of reports on scientific assessment of ozone depletion. The most recent is from 2007 where satellite measurements have shown the hole in the ozone layer is recovering and is now the smallest it has been for about a decade.

Ozone Depletion and Global Warming

Although they are often interlinked in the mass media, the connection between global warming and ozone depletion is not strong. There are five areas of linkage:

- The same CO_2 radiative forcing that produces near-surface global warming is expected to cool the stratosphere. This cooling, in turn, is expected to produce a relative *increase* in polar ozone (O_3) depletion and the frequency of ozone holes.
- Conversely, ozone depletion represents a radiative forcing of the climate system. There are two opposing effects: Reduced ozone causes the stratosphere to absorb less solar radiation, thus cooling the stratosphere while warming the troposphere; the resulting colder stratosphere emits less long-wave radiation downward, thus cooling the troposphere. Overall, the cooling dominates; the IPCC concludes that "*observed stratospheric O_3 losses over the past two decades have caused a negative forcing of the surface-troposphere system*" of about "0.15 ± 0.10 watts per square meter (W/m^2).
- One of the strongest predictions of the greenhouse effect is that the stratosphere will cool. Although this cooling

has been observed, it is not trivial to separate the effects of changes in the concentration of greenhouse gases and ozone depletion since both will lead to cooling. However, this can be done by numerical stratospheric modeling. Results from the National Oceanic and Atmospheric Administration's Geophysical Fluid Dynamics Laboratory show that above 20 km (12.4 miles), the greenhouse gases dominate the cooling.

- Ozone depleting chemicals are also greenhouse gases. The increases in concentrations of these chemicals have produced 0.34 ± 0.03 W/m^2 of radiative forcing, corresponding to about 14% of the total radiative forcing from increases in the concentrations of well-mixed greenhouse gases.
- The long-term modeling of the process, its measurement, study, design of theories and testing take decades to both document, gain wide acceptance, and ultimately become the dominant paradigm. Several theories about the destruction of ozone, were hypothesized in the 1980s, published in the late 1990s, and are currently being proven. Dr Drew Schindell, and Dr Paul Newman, NASA Goddard, proposed a theory in the late 1990s, using a SGI Origin 2000 supercomputer, that modeled ozone destruction, accounted for 78% of the ozone destroyed. Further refinement of that model, accounted for 89% of the ozone destroyed, but pushed back the estimated recovery of the ozone hole from 75 years to 150 years. (An important part of that model is the lack of stratospheric flight due to depletion of fossil fuels.)

Misconceptions about Ozone Depletion

CFCs are "Too Heavy" to reach the Stratosphere

It is sometimes stated that since CFC molecules are much heavier than nitrogen or oxygen, they cannot reach the stratosphere in significant quantities. But atmospheric gases are not sorted by weight; the forces of wind (turbulence) are strong enough to fully intermix gases in the atmosphere. CFCs are heavier than air, but just like argon, krypton and other heavy gases with a long lifetime, they are uniformly distributed throughout the turbosphere and reach the upper atmosphere.

Man-made Chlorine is Insignificant Compared to Natural Sources

Another objection occasionally voiced is that *It is generally agreed that natural sources of tropospheric chlorine (volcanoes, ocean spray, etc.) are four to five orders of magnitude larger than man-made sources*. While strictly true, *tropospheric* chlorine is irrelevant; it is *stratospheric* chlorine that affects ozone depletion. Chlorine from ocean spray is soluble and thus is washed out by rainfall before it reaches the stratosphere. CFCs, in contrast, are insoluble and long-lived, which allows them to reach the stratosphere. Even in the lower atmosphere there is more chlorine present in the form of CFCs and related haloalkanes than there is in HCl from salt spray, and in the stratosphere halocarbons dominate overwhelmingly. Only one of these halocarbons, methyl chloride, has a predominantly natural source, and it is responsible for about 20 per cent of the chlorine in the stratosphere; the remaining 80% comes from manmade compounds.

Very large volcanic eruptions can inject HCl directly into the stratosphere, but direct measurements have shown that their contribution is small compared to that of chlorine from CFCs. A similar erroneous assertion is that soluble halogen compounds from the volcanic plume of Mount Erebus on Ross Island, Antarctica are a major contributor to the Antarctic ozone hole.

An Ozone Hole was First Observed in 1956

G.M.B. Dobson (Exploring the Atmosphere, 2nd Edition, Oxford, 1968) mentioned that when springtime ozone levels over Halley Bay were first measured, he was surprised to find that they were ~320 DU, about 150 DU below spring levels, ~450 DU, in the Arctic. These, however, were the pre-ozone hole normal climatological values. What Dobson describes is essentially the *baseline* from which the ozone hole is measured: actual ozone hole values are in the 150–100 DU range.

The discrepancy between the Arctic and Antarctic noted by Dobson was primarily a matter of timing: during the Arctic spring ozone levels rose smoothly, peaking in April, whereas in the Antarctic they stayed approximately constant during early spring, rising abruptly in November when the polar vortex broke down.

The behaviour seen in the Antarctic ozone hole is completely different. Instead of staying constant, early springtime ozone levels suddenly drop from their already low winter values, by as much as 50%, and normal values are not reached again until December.

If the Theory were correct, the Ozone Hole should be above the Sources of CFCs

CFCs are well mixed in the troposphere and the stratosphere. The reason the ozone hole occurs above Antarctica is not because there are more CFCs there but because the low temperatures allow polar stratospheric clouds to form. There have been anomalous discoveries of significant, serious, localized "holes" above other parts of the globe.

The "Ozone Hole" is a Hole in the Ozone Layer

When the "ozone hole" forms, essentially all of the ozone in the lower stratosphere is destroyed. The upper stratosphere is much less affected, however, so that the overall amount of ozone over the continent declines by 50 per cent or more. The ozone hole does not go all the way through the layer; on the other hand, it is not a uniform 'thinning' of the layer either. It is a "hole" in the sense of "a hole in the ground", that is, a depression; not in the sense of "a hole in the windshield."

World Ozone Day

In 1994, the United Nations General Assembly voted to designate 16 September as "World Ozone Day", to commemorate the signing of the Montreal Protocol on that date in 1987.

12

The Environmental Protection Act 1986

The Environment (Protection) Act, 1986

No. 29 of 1986

An Act to provide for the protection and improvement of environment and for matters connected there with:

WHEREAS the decisions were taken at the United Nations Conference on the Human Environment held at Stockholm in June, 1972, in which India participated, to take appropriate steps for the protection and improvement of human environment;

AND WHEREAS it is considered necessary further to implement the decisions aforesaid in so far as they relate to the protection and improvement of environment and the prevention of hazards to human beings, other living creatures, plants and property;

BE it enacted by Parliament in the Thirty-seventh Year of the Republic of India as follows:

CHAPTER I

PRELIMINARY

1. Short Title, Extend and Commencement

(1) This Act may be called the Environment (Protection) Act, 1986.

(2) It extends to the whole of India.

(3) It shall come into force on such date as the Central Government may, by notification in the Official Gazette, appoint and different dates may be appointed for different provisions of this Act and for different areas.

2. Definitions

In this Act, unless the context otherwise requires,—

(a) "environment" includes water, air and land and the inter- relationship which exists among and between water, air and land, and human beings, other living creatures, plants, micro-organism and property;

(b) "environmental pollutant" means any solid, liquid or gaseous substance present in such concentration as may be, or tend to be, injurious to environment;

(c) "environmental pollution" means the presence in the environment of any environmental pollutant;

(d) "handling", in relation to any substance, means the manufacture, processing, treatment, package, storage, transportation, use, collection, destruction, conversion, offering for sale, transfer or the like of such substance;

(e) "hazardous substance" means any substance or preparation which, by reason of its chemical or physico-chemical properties or handling, is liable to cause harm to human beings, other living creatures, plant, micro-organism, property or the environment;

(f) "occupier", in relation to any factory or premises, means a person who has, control over the affairs of the factory or the premises and includes in relation to any substance, the person in possession of the substance;

(*g*) "prescribed" means prescribed by rules made under this Act.

CHAPTER II

GENERAL POWERS OF THE CENTRAL GOVERNMENT

3. Power of Central Government to take Measures to Protect and Improve Environment

(1) Subject to the provisions of this Act, the Central Government, shall have the power to take all such measures as it deems necessary or expedient for the purpose of protecting and improving the quality of the environment and preventing controlling and abating environmental pollution.

(2) In particular, and without prejudice to the generality of the provisions of sub-section (1), such measures may include measures with respect to all or any of the following matters, namely:

(*i*) co-ordination of actions by the State Governments, officers and other authorities—

(*a*) under this Act, or the rules made thereunder, or

(*b*) under any other law for the time being in force which is relatable to the objects of this Act;

(*ii*) planning and execution of a nation-wide programme for the prevention, control and abatement of environmental pollution;

(*iii*) laying down standards for the quality of environment in its various aspects;

(*iv*) laying down standards for emission or discharge of environmental pollutants from various sources whatsoever:

Provided that different standards for emission or discharge may be laid down under this clause from different sources having regard to the quality or composition of the emission or discharge of environmental pollutants from such sources;

(v) restriction of areas in which any industries, operations or processes or class of industries, operations or processes shall not be carried out or shall be carried out subject to certain safeguards;

(vi) laying down procedures and safeguards for the prevention of accidents which may cause environmental pollution and remedial measures for such accidents;

(vii) laying down procedures and safeguards for the handling of hazardous substances;

(viii) examination of such manufacturing processes, materials and substances as are likely to cause environmental pollution;

(ix) carrying out and sponsoring investigations and research relating to problems of environmental pollution;

(x) inspection of any premises, plant, equipment, machinery, manufacturing or other processes, materials or substances and giving, by order, of such directions to such authorities, officers or persons as it may consider necessary to take steps for the prevention, control and abatement of environmental pollution;

(xi) establishment or recognition of environmental laboratories and institutes to carry out the functions entrusted to such environmental laboratories and institutes under this Act;

(*xii*) collection and dissemination of information in respect of matters relating to environmental pollution;

(*xiii*) preparation of manuals, codes or guides relating to the prevention, control and abatement of environmental pollution;

(*xiv*) such other matters as the Central Government deems necessary or expedient for the purpose of securing the effective implementation of the provisions of this Act.

(3) The Central Government may, if it considers it necessary or expedient so to do for the purpose of this Act, by order, published in the Official Gazette, constitute an authority or authorities by such name or names as may be specified in the order for the purpose of exercising and performing such of the powers and functions (including the power to issue directions under section 5) of the Central Government under this Act and for taking measures with respect to such of the matters referred to in sub-section (2) as may be mentioned in the order and subject to the supervision and control of the Central Government and the provisions of such order, such authority or authorities may exercise and powers or perform the functions or take the measures so mentioned in the order as if such authority or authorities had been empowered by this Act to exercise those powers or perform those functions or take such measures.

4. Appointment of Officers and their Powers and Functions

(1) Without prejudice to the provisions of sub-section (3) of section 3, the Central Government may appoint officers with such designation as it thinks fit for the purposes of this Act and may entrust to them such of

the powers and functions under this Act as it may deem fit.

(2) The officers appointed under sub-section (1) shall be subject to the general control and direction of the Central Government or, if so directed by that Government, also of the authority or authorities, if any, constituted under sub- section (3) of section 3 or of any other authority or officer.

5. Power to Give Directions

Notwithstanding anything contained in any other law but subject to the provisions of this Act, the Central Government may, in the exercise of its powers and performance of its functions under this Act, issue directions in writing to any person, officer or any authority and such person, officer or authority shall be bound to comply with such directions.

Explanation—For the avoidance of doubts, it is hereby declared that the power to issue directions under this section includes the power to direct—

(a) the closure, prohibition or regulation of any industry, operation or process; or

(b) stoppage or regulation of the supply of electricity or water or any other service.

6. Rules to Regulate Environmental Pollution

(1) The Central Government may, by notification in the Official Gazette, make rules in respect of all or any of the matters referred to in section 3.

(2) In particular, and without prejudice to the generality of the foregoing power, such rules may provide for all or any of the following matters, namely:

(a) the standards of quality of air, water or soil for various areas and purposes;

(*b*) the maximum allowable limits of concentration of various environmental pollutants (including noise) for different areas;

(*c*) the procedures and safeguards for the handling of hazardous substances;

(*d*) the prohibition and restrictions on the handling of hazardous substances in different areas;

(*e*) the prohibition and restriction on the location of industries and the carrying on process and operations in different areas;

(*f*) the procedures and safeguards for the prevention of accidents which may cause environmental pollution and for providing for remedial measures for such accidents.

CHAPTER III

PREVENTION, CONTROL, AND ABATEMENT OF ENVIRONMENTAL POLLUTION

7. Persons carrying on Industry Operation, etc., not to allow Emission or Discharge of Environmental Pollutants in Excess of the Standards

No person carrying on any industry, operation or process shall discharge or emit or permit to be discharged or emitted any environmental pollutants in excess of such standards as may be prescribed.

8. Persons Handling Hazardous Substances to comply with Procedural Safeguards

No person shall handle or cause to be handled any hazardous substance except in accordance with such procedure and after complying with such safeguards as may be prescribed.

9. Furnishing of Information to Authorities and Agencies in Certain Cases

(1) Where the discharge of any environmental pollutant in excess of the prescribed standards occurs or is apprehended to occur due to any accident or other unforeseen act or event, the person responsible for such discharge and the person in charge of the place at which such discharge occurs or is apprehended to occur shall be bound to prevent or mitigate the environmental pollution caused as a result of such discharge and shall also forthwith—

(a) intimate the fact of such occurrence or apprehension of such occurrence; and

(b) be bound, if called upon, to render all assistance, to such authorities or agencies as may be prescribed.

(2) On receipt of information with respect to the fact or apprehension on any occurrence of the nature referred to in sub-section (1), whether through intimation under that sub-section or otherwise, the authorities or agencies referred to in sub-section (1) shall, as early as practicable, cause such remedial measures to be taken as necessary to prevent or mitigate the environmental pollution.

(3) The expenses, if any, incurred by any authority or agency with respect to the remedial measures referred to in sub-section (2), together with interest (at such reasonable rate as the Government may, by order, fix) from the date when a demand for the expenses is made until it is paid, may be recovered by such authority or agency from the person concerned as arrears of land revenue or of public demand.

10. Powers of Entry and Inspection

(1) Subject to the provisions of this section, any person empowered by the Central Government in this behalf shall have a right to enter, at all reasonable times with such assistance as he considers necessary, any place—

(a) for the purpose of performing any of the functions of the Central Government entrusted to him;

(b) for the purpose of determining whether and if so in what manner, any such functions are to be performed or whether any provisions of this Act or the rules made thereunder or any notice, order, direction or authorisation served, made, given or granted under this Act is being or has been complied with;

(c) for the purpose of examining and testing any equipment, industrial plant, record, register, document or any other material object or for conducting a search of any building in which he has reason to believe that an offence under this Act or the rules made thereunder has been or is being or is about to be committed and for seizing any such equipment, industrial plant, record, register, document or other material object if he has reason to believe that it may furnish evidence of the commission of an offence punishable under this Act or the rules made thereunder or that such seizure is necessary to prevent or mitigate environmental pollution.

(2) Every person carrying on any industry, operation or process of handling any hazardous substance shall be bound to render all assistance to the person empowered by the Central Government under sub-section (1) for carrying out the functions under that

sub-section and if he fails to do so without any reasonable cause or excuse, he shall be guilty of an offence under this Act.

(3) If any person wilfully delays or obstructs any persons empowered by the Central Government under sub-section (1) in the performance of his functions, he shall be guilty of an offence under this Act.

(4) The provisions of the Code of Criminal Procedure, 1973, or, in relation to the State of Jammu and Kashmir, or an area in which that Code is not in force, the provisions of any corresponding law in force in that State or area shall, so far as may be, apply to any search or seizures under this section as they apply to any search or seizure made under the authority of a warrant issued under section 94 of the said Code or as the case may be, under the corresponding provision of the said law.

11. Power to take Sample and Procedure to be followed in Connection therewith

(1) The Central Government or any officer empowered by it in this behalf, shall have power to take, for the purpose of analysis, samples of air, water, soil or other substance from any factory, premises or other place in such manner as may be prescribed.

(2) The result of any analysis of a sample taken under sub-section (1) shall not be admissible in evidence in any legal proceeding unless the provisions of sub-sections (3) and (4) are complied with.

(3) Subject to the provisions of sub-section (4), the person taking the sample under sub-section (1) shall—

(a) serve on the occupier or his agent or person in charge of the place, a notice, then and there, in

such form as may be prescribed, of his intention to have it so analysed;

(b) in the presence of the occupier of his agent or person, collect a sample for analysis;

(c) cause the sample to be placed in a container or containers which shall be marked and sealed and shall also be signed both by the person taking the sample and the occupier or his agent or person;

(d) send without delay, the container or the containers to the laboratory established or recognized by the Central Government under section 12.

(4) When a sample is taken for analysis under sub-section (1) and the person taking the sample serves on the occupier or his agent or person, a notice under clause (a) of sub-section (3), then,—

(a) in a case where the occupier, his agent or person wilfully absents himself, the person taking the sample shall collect the sample for analysis to be placed in a container or containers which shall be marked and sealed and shall also be signed by the person taking the sample, and

(b) in a case where the occupier or his agent or person present at the time of taking the sample refuses to sign the marked and sealed container or containers of the sample as required under clause (c) of sub-section (3), the marked and sealed container or containers shall be signed by the person taking the samples, and the container or containers shall be sent without delay by the person taking the sample for analysis to the laboratory established or recognized under section 12 and such person shall inform the Government Analyst appointed

or recognised under section 12 in writing, about the wilfull absence of the occupier or his agent or person, or, as the case may be, his refusal to sign the container or containers.

12. Environmental Laboratories

(1) The Central Government may, by notification in the Official Gazette,—

(*a*) establish one or more environmental laboratories;

(*b*) recognise one or more laboratories or institutes as environmental laboratories to carry out the functions entrusted to an environmental laboratory under this Act.

(2) The Central Government may, by notification in the Official Gazette, make rules specifying—

(*a*) the functions of the environmental laboratory;

(*b*) the procedure for the submission to the said laboratory of samples of air, water, soil or other substance for analysis or tests, the form of the laboratory report thereon and the fees payable for such report;

(*c*) such other matters as may be necessary or expedient to enable that laboratory to carry out its functions.

13. Government Analysts

The Central Government may by notification in the Official Gazette, appoint or recognise such persons as it thinks fit and having the prescribed qualifications to be Government Analysts for the purpose of analysis of samples of air, water, soil or other substance sent for analysis to any environmental laboratory established or recognised under sub-section (1) of section 12.

14. Reports of Government Analysts

Any document purporting to be a report signed by a Government analyst may be used as evidence of the facts stated therein in any proceeding under this Act.

15. Penalty for Contravention of the Provisions of the Act and the Rules, Orders and Directions

(1) Whoever fails to comply with or contravenes any of the provisions of this Act, or the rules made or orders or directions issued thereunder, shall, in respect of each such failure or contravention, be punishable with imprisonment for a term which may extend to five years with fine which may extend to one lakh rupees, or with both, and in case the failure or contravention continues, with additional fine which may extend to five thousand rupees for every day during which such failure or contravention continues after the conviction for the first such failure or contravention.

(2) If the failure or contravention referred to in sub-section (1) continues beyond a period of one year after the date of conviction, the offender shall be punishable with imprisonment for a term which may extend to seven years.

16. Offences by Companies

(1) Where any offence under this Act has been committed by a company, every person who, at the time the offence was committed, was directly in charge of, and was responsible to, the company for the conduct of the business of the company, as well as the company, shall be deemed to be guilty of the offence and shall be liable to be proceeded against and punished accordingly:

Provided that nothing contained in this sub-section shall render any such person liable to any punishment provided in this Act, if he proves that the offence was committed without his knowledge or that he exercised all due diligence to prevent the commission of such offence.

(2) Notwithstanding anything contained in sub-section (1), where an offence under this Act has been committed by a company and it is proved that the offence has been committed with the consent or connivance of, or is attributable to any neglect on the part of, any director, manager, secretary or other officer of the company, such director, manager, secretary or other officer shall also deemed to be guilty of that offence and shall be liable to be proceeded against and punished accordingly.

Explanation—For the purpose of this section,—

(a) "company" means any body corporate and includes a firm or other association of individuals;

(b) "director", in relation to a firm, means a partner in the firm.

17. Offences by Government Departments

(1) Where an offence under this Act has been committed by any Department of Government, the Head of the Department shall be deemed to be guilty of the offence and shall be liable to be proceeded against and punished accordingly.

Provided that nothing contained in this section shall render such Head of the Department liable to any punishment if he proves that the offence was committed without his knowledge or that he exercise all due diligence to prevent the commission of such offence.

(2) Notwithstanding anything contained in sub-section (1), where an offence under this Act has been

committed by a Department of Government and it is proved that the offence has been committed with the consent or connivance of, or is attributable to any neglect on the part of, any officer, other than the Head of the Department, such officer shall also be deemed to be guilty of that offence and shall be liable to be proceeded against and punished accordingly.

CHAPTER IV

MISCELLANEOUS

18. Protection of Action taken in Good Faith

No suit, prosecution or other legal proceeding shall lie against the Government or any officer or other employee of the Government or any authority constituted under this Act or any member, officer or other employee of such authority in respect of anything which is done or intended to be done in good faith in pursuance of this Act or the rules made or orders or directions issued thereunder.

19. Cognizance of Offences

No court shall take cognizance of any offence under this Act except on a complaint made by—

(*a*) the Central Government or any authority or officer authorised in this behalf by that Government, or

(*b*) any person who has given notice of not less than sixty days, in the manner prescribed, of the alleged offence and of his intention to make a complaint, to the Central Government or the authority or officer authorised as aforesaid.

20. Information, Reports or Returns

The Central Government may, in relation to its function under this Act, from time to time, require any person, officer,

State Government or other authority to furnish to it or any prescribed authority or officer any reports, returns, statistics, accounts and other information and such person, officer, State Government or other authority shall be bound to do so.

21. Members, Officers and Employees of The Authority Constituted under section 3 to be Public Servants

All the members of the authority, constituted, if any, under section 3 and all officers and other employees of such authority when acting or purporting to act in pursuance of any provisions of this Act or the rules made or orders or directions issued thereunder shall be deemed to be public servants within the meaning of section 21 of the Indian Penal Code (45 of 1860).

22. Bar of Jurisdiction

No civil court shall have jurisdiction to entertain any suit or proceeding in respect of anything done, action taken or order or direction issued by the Central Government or any other authority or officer in pursuance of any power conferred by or in relation to its or his functions under this Act.

23. Powers to Delegate

Without prejudice to the provisions of sub-section (3) of section 3, the Central Government may, by notification in the Official Gazette, delegate, subject to such conditions and limitations as may be specified in the notifications, such of its powers and functions under this Act [except the powers to constitute an authority under sub-section (3) of section 3 and to make rules under section 25] as it may deem necessary or expedient, to any officer, State Government or other authority.

24. Effect of Other Laws

(1) Subject to the provisions of sub-section (2), the provisions of this Act and the rules or orders made

therein shall have effect notwithstanding anything inconsistent therewith contained in any enactment other than this Act.

(2) Where any act or omission constitutes an offence punishable under this Act and also under any other Act then the offender found guilty of such offence shall be liable to be punished under the other Act and not under this Act.

25. Power to Make Rules

(1) The Central Government may, by notification in the Official Gazette, make rules for carrying out the purposes of this Act.

(2) In particular, and without prejudice to the generality of the foregoing power, such rules may provide for all or any of the following matters, namely—

(a) the standards in excess of which environmental pollutants shall not be discharged or emitted under section 7;

(b) the procedure in accordance with and the safeguards in compliance with which hazardous substances shall be handled or caused to be handled under section 8;

(c) the authorities or agencies to which intimation of the fact of occurrence or apprehension of occurrence of the discharge of any environmental pollutant in excess of the prescribed standards shall be given and to whom all assistance shall be bound to be rendered under sub-section (1) of section 9;

(d) the manner in which samples of air, water, soil or other substance for the purpose of analysis shall be taken under sub-section (1) of section 11;

(*e*) the form in which notice of intention to have a sample analysed shall be served under clause (a) of sub section (3) of section 11;

(*f*) the functions of the environmental laboratories, the procedure for the submission to such laboratories of samples of air, water, soil and other substances for analysis or test; the form of laboratory report; the fees payable for such report and other matters to enable such laboratories to carry out their functions under sub-section (2) of section 12;

(*g*) the qualifications of Government Analyst appointed or recognised for the purpose of analysis of samples of air, water, soil or other substances under section 13;

(*h*) the manner in which notice of the offence and of the intention to make a complaint to the Central Government shall be given under clause (b) of section 19;

(*i*) the authority of officer to whom any reports, returns, statistics, accounts and other information shall be furnished under section 20;

(*j*) any other matter which is required to be, or may be, prescribed.

26. Rules made under this Act to be laid before Parliament

Every rule made under this Act shall be laid, as soon as may be after it is made, before each House of Parliament, while it is in session, for a total period of thirty days which may be comprised in one session or in two or more successive sessions, and if, before the expiry of the session immediately following the session or the successive sessions aforesaid, both Houses agree in making any modification in the rule or both Houses agree that the rule should not be made, the rule shall thereafter have effect only in such modified form or be of no effect, as the case may be; so, however, that any such modification or annulment shall be without prejudice to the validity of anything previously done under that rule.

Index

R

S

T

V

W